Student Success Guide

TO ACCOMPANY

WHAT IS LIFE? A GUIDE TO BIOLOGY

Second Edition

Meredith S. Norris, M.S.
Jennifer M. Warner, Ph.D.
Department of Biology
The University of North Carolina at Charlotte

W.H. Freeman and Company
New York

ISBN-13: 9781464106774
ISBN-10: 1464106770

Visual Glossary adapted by Max Pepper.

Printed in the United States of America

First printing

W.H. Freeman and Company
41 Madison Avenue
New York, NY 10010
Houndmills, Basingstoke
RG21 6XS, England

www.whfreeman.com

CONTENTS

ABOUT THE AUTHORS

Meredith S. Norris is a lecturer on the faculty of the Department of Biology at the University of North Carolina at Charlotte. She received her M.S. degree in Immunology with a focus on large animal medicine from the University of California at Davis following her undergraduate career as an animal science major at the University of Massachusetts at Amherst. Meredith has taught a variety of laboratory courses in animal science, equine science and biological sciences, but currently focuses on introductory biology in large lectures for non-science majors as well as general biology for science majors. She has also taught online general education courses that explore contemporary issues related to science and society.

Meredith is a member of the National Association of Biology Teachers and her pedagogical interests include incorporating technology in the classroom and active learning. Meredith currently lives in Charlotte, North Carolina, with her husband and son.

Dr. Jennifer Warner is the Vice Chair of Academic Programs in the Department of Biology at the University of North Carolina at Charlotte and a National Academies Education Fellow. She received her M.S. degree in Microbiology and her Ph.D. in Curriculum and Instruction with concentrations in science education and higher education administration. Jennifer has taught more than 40 sections of non-majors biology. She has also taught classes in human biology, anatomy and physiology, microbiology, pathogenic bacteriology, , and the nature of science. She frequently leads teacher training workshops for middle school and high school teachers and has authored test preparation books for the MCAT and PCAT.

Jennifer's research interests center around equity and gender issues in science education, Supplemental Instruction (SI) programming, and the role of science identity in transitions of college students from non-science to science majors. Jennifer resides in Charlotte, North Carolina with her husband, daughter, and two extraordinarily spoiled hounds.

PREFACE

A Note to Our Colleagues

The inspiration for this guide came from many years of working with non-majors students and finding dissatisfaction in the study resources available for them. This guide is based on a model that has been used successfully for thirteen years with more than 15,000 students at the University of North Carolina at Charlotte. It is our opinion that this guide is so different from other study guides that we prefer not to call it a study guide and instead refer to it as a student success guide.

The problems we see with many study guides is that they are too lengthy and do not emphasize concepts the way many of us test on concepts. The result is that students become frustrated and do not use these guides. This guide was developed to be brief and to the point. Students work through the chapter concepts using an outline and then test themselves using multiple choice and short answer questions. They are also provided with a visual listing of key terms to help them focus in on critical chapter concepts.

There are two ways you might choose to use this guide. One way is to have the students complete the outlines during your class lectures as a way to take notes. Space has been left in the margin for students to add additional notes during class. If you prefer not to use the outlines during class, students can complete them on their own providing them with a great review of the chapter. After the students have completed the outlines they can use the Testing and Applying Your Understanding questions. Each chapter has a reasonable number of multiple choice and short answer questions to adequately test students' knowledge but not so many questions that the student is overwhelmed. The questions provided are meant to be more challenging so that the student can be confident that they truly understand concepts once they have completed the work in this guide.

Our intention with this guide is to provide students with a study plan to help them achieve success in their biology course. The guide is meant to provide a seamless connection between your class lectures, the text chapters, and the study materials. Our experience with our own students is that they frequently comment that the structure of this guide is a primary factor in their success in our courses. We hope you will hear the same comments from your students.

To the Student: How to Use this Guide

You have probably already formed an opinion about most of the study guides you have used in the past. Very often those guides require more time than you might have to complete and they may not test you in the same way that your professor will on a test. For these reasons, you may be skeptical about using this study guide. We hope that you will give this guide a chance. We think it is so different that we aren't even calling it a study guide – we prefer the term student success guide. This guide was written to be brief, to the point, and to test you in a way that is similar to what you can expect from your professor. We understand the material can be challenging and at times intimidating but we feel confident that the time you invest to complete this guide will help you be successful in your biology class.

Here is the best way to use this guide:

1. It is always best to read the text chapter before you attend your lecture class, so make sure you schedule time for this important task. It is also a great idea to review the **Visual Learning Glossary** in this guide before you attend class.

2. After reading the text chapter, you are ready to complete the **Chapter Outline**. Depending on your professor's preferences, you may fill out the outline during class lecture. There is space in the margins for you to add in any extra notes during class. If your professor is not using the outlines in class, you should complete them on your own. Filling in these outlines helps organize the concepts and helps you retain the material.

3. After reading the text chapter and completing the outline for each chapter, work on the **Testing and Applying Your Understanding** questions. You will find multiple choice and short answer questions in each chapter. Some of these questions are meant to be challenging because we want you to be prepared for the sorts of questions your professor might ask you on an exam. You will also notice that some of the questions ask you to think critically or apply material to a unique situation. Answering these sorts of questions is an important test of your true understanding of a concept.

4. Check your answers to the questions for the guide. You can find answers to the multiple choice questions at the end of the book and answers to the short answer questions on the Phelan student Web site at www.whfreeman.com/phelan2e.

5. Timing is everything! You should complete the resources for each chapter in this guide as soon as possible after your lecture. If you wait too long, the work builds up and you begin forgetting details. Completing each chapter within a few days of covering the topic in class will help the material make its way to your long term memory. It also ensures that you pace yourself and won't find yourself with a pile of work the day or two before an exam.

6. Exam time is approaching . . . If you have completed the guide for each chapter in a timely manner, you will be well on your way to preparedness for your exam. Now is the time to review the chapters, your class notes, the outlines, the key terms and the questions for each chapter. Don't forget that hundreds of additional multiple-choice questions are available at Prep-U: www.prep-u.com.

We wish you the best of luck in your biology class. If you follow the suggestions provided by your professor and in this guide, you will be well on your way to success. We hope you will enjoy learning about biology and cultivate an appreciation for just how much biology impacts your everyday life!

Chapter 1
SCIENTIFIC THINKING—YOUR BEST PATHWAY TO UNDERSTANDING THE WORLD

Learning Objectives

- Differentiate between scientific and alternative ways of thinking
- Describe the attributes of scientific and biological literacy
- Apply the scientific method to everyday situations
- Understand the flexibility of the scientific method
- Design critical experiments to test hypotheses
- Understand the importance of well-designed and controlled experiments
- Describe ways to blind experiments
- Describe the biases that can occur in scientific studies and ways to reduce those biases
- Understand how statistics can add support to scientific claims
- Identify pseudoscience and anecdotal evidence masked as scientific evidence
- Understand the limitations of science as a means to answer questions
- Understand how hierarchical organization and evolution tie together all the major themes in biology

Chapter Outline

I. Science is a Collection of Facts and a Process for Understanding the World

- Define **science**:

- Define **biology**:

- Define **scientific literacy**:

- The development of **biological literacy** is important for understanding social, political, medical, and legal issues. This involves the ability to:

  1.

  2.

  3.

- Explain how **superstition** is different from scientific thinking?

- Describe **empirical** knowledge:

- Provide three examples of non-scientific ways of thinking.

    1.

    2.

    3.

II. A Beginner's Guide: What Are the Steps of the Scientific Method?

THE SCIENTIFIC METHOD

STEP 1   STEP 2   STEP 3   STEP 4   STEP 5

- While some people envision scientists in labs checking off the steps of the **scientific method** as they work, in reality the scientific method is more of a flexible set of guidelines for scientific thinking. The basic steps in the method include:

    1.

    2.

    3.

    4.

    5.

- Describe what is meant by the term self-correcting in reference to the scientific method:

- A _____ is defined as tentative explanation for observed phenomena.

- A useful **hypothesis** must be able to achieve two goals:

  1.

  2.

- A hypothesis that is stated such that there is not a relationship between two variables is referred to as a _____ hypothesis.

- In terms of hypotheses, it is best to state that they can be _____ by data instead of proven.

- Useful hypotheses involve testable predictions. Hypotheses can be refined and adjusted as needed.

- An experiment that makes it possible to support or disprove a hypothesis is called a _____ experiment.

- Explain how a **placebo** is used in a critical experiment.

- Hypotheses with overwhelming experimental support can evolve into theories. While the common use of the term **theory** refers to a hunch or tentative hypothesis, the scientific use of theory means something much more concrete with extensive empirical evidence to support the theory.

III. Well-Designed Experiments Are Essential to Testing Hypotheses

- Experimental design is a critical part of the scientific process. Experiments that are carefully and purposefully designed are most likely to show cause-and-effect relationships between variables. Additionally, it is critical that experimental results can be reproduced when the experiment is repeated.

A. Controlled Experiments

- Define each of the following elements of experiments:

    1. Treatment

    2. Experimental group

    3. Control group

    4. Variables

- Why is it so important to control variables in an experiment?

- The use of placebos can lead to the **placebo effect**. Explain what this means.

B. Blinding

- Blinding is a way to reduce bias in reporting and analyzing data. In a **blind experimental design**, the _____ are not aware whether they are receiving the treatment or not. In a **double-blind experimental design**, neither the _____ nor the _____ are aware of which subjects are receiving the treatment.

- Blinded experimental designs can also been improved by **randomizing** the study. What does this mean?

IV. The Scientific Method Can Help Us Make Wise Decisions

- The use of the scientific method and a critical approach to reviewing evidence can help us make sense of data and anecdotal evidence, as well as help us distinguish science from pseudoscience.

A. Visual Displays of Data

- Which forms of displaying data are used most frequently in biology?

- Describe each of the common elements found in most visual displays of data:

    1. Title

    2. *x*-axis

    3. *y*-axis

    4. Independent variable

    5. Dependent variable

    6. Data points

B. Statistics

- How can **statistics** put data in context to determine if a hypothesis is supported or not?

- When evaluating statistical analysis of data in order to determine whether the treatment had a significant effect or not, we are looking for a _____ difference between the experimental and control groups with a _____ variation within each group.

- **Correlational studies** look at relationships between different variables. A _____ correlation is when an increase in one variable leads to an increase in another variable. A _____ correlation occurs when an increase in one variable leads to a decrease in another variable.

C. Pseudoscience and Anecdotal Evidence

- People are often taken in by false scientific claims. The two most common ways to mislead the public are by using pseudoscience and anecdotal evidence. Describe what these terms mean:

  o Pseudoscience

  o Anecdotal evidence

- Describe an example of misleading scientific evidence in the form of pseudoscience and anecdotal evidence:

  o Pseudoscience

  o Anecdotal evidence

D. The Limits of Science

- While we often talk about what science is, we don't always hear so much about what science is not. Science, just like other disciplines, has limitations to the sorts of questions it can answer.

- Give several examples of questions that science cannot answer:

V.   On the Road to Biological Literacy: What Are the Major Themes in Biology?

- As you progress through the course you will learn about many differing aspects of biology. However, there are two common themes that will prevail throughout the course. Explain these major themes:

  o Hierarchical organization

  o Evolution

Testing and Applying Your Understanding

Multiple Choice (For more multiple choice questions, visit www.prep-u.com.)

1. In controlled experiments:
   a) one variable is manipulated while others are held constant.
   b) all variables are held constant.
   c) all variables are dependent on each other.
   d) all variables are independent of each other.
   e) all critical variables are manipulated.

2. In the late 1950s, a doctor reported in the *Journal of the American Medical Association* that stomach ulcers could be effectively treated by having a patient swallow a balloon connected to some tubes that circulated a refrigerated fluid. He argued that, by super-cooling the stomach, acid production was reduced and the ulcer relieved. All 24 of his patients who received the treatment were healed. Why does this fall short of qualifying as an example of the scientific method?
   a) All of the patients were aware of the treatment they received.
   b) There was no control group with which to compare his patients who received the treatment.
   c) Although there was a control group, it was not randomly selected.
   d) There were not enough experimental subjects to draw a definitive conclusion.
   e) All of the above are correct.

3. The placebo effect:
   a) is an urban legend.
   b) reveals that sugar pills are generally as effective as actual medications in fighting illness.
   c) demonstrates that most scientific studies cannot be replicated.
   d) is the frequently observed, poorly understood phenomenon that people tend to respond favorably to any treatment.
   e) reveals that experimental treatments cannot be proven as effective.

4. In many reptiles, the sex of a fetus is determined by the incubation temperature of the egg; higher temperatures lead to more males. However, DDE (a chemical byproduct of DDT) in the environment prior to birth drastically lowers the normal percentage of males. You want to design a good scientific experiment to illustrate this phenomenon, but before you can, you must properly identify the different components of the experiment. Which of the following choices does not properly identify each of the experimental components in this example?
   a) The sample size of your experiment would be the number of eggs you test on.
   b) The constant in this experiment would be the application of DDE.
   c) The experimental group would be the group of eggs you did expose to DDE prior to birth.
   d) The control group would be the group of eggs you did not expose to DDE prior to birth.
   e) All of the above are correct.

5. By 1796 it had been observed that milkmaids who had been exposed to cowpox did not succumb to the deadly plague of small pox that was ravishing both Europe and Britain. From this observation, Edward Jenner was able to construct the world's first successful vaccine. If you had to predict the hypothesis that Jenner made to lead him to his creation, which of the following choices best fits your prediction?
    a) If milkmaids exposed to cowpox are immune to smallpox, then cowpox and smallpox are the same disease.
    b) If exposure to cowpox gives immunity to smallpox in milkmaids, then milkmaids have a natural immunity and their blood should be used to develop a smallpox vaccine.
    c) If exposure to cowpox gives immunity to smallpox in milkmaids, then exposure to cowpox should give immunity to smallpox in other individuals as well.
    d) Cowpox and smallpox are caused by the same microorganism.
    e) Exposure to smallpox always gives immunity to cowpox.

6. Science as a way of seeking principles of order differs from art, religion, and philosophy in that:
    a) all scientific knowledge is gained by experimentation.
    b) science limits its search to the natural world of the physical universe.
    c) there is no room for intuition or guessing.
    d) science denies the existence of the supernatural.
    e) science deals exclusively with known facts.

7. Experimental drugs must undergo many rigorous trials to ensure they deliver their medical benefits effectively and safely. One method that is commonly used in this process is to compare the effects of a drug to that of a neutral placebo in double-blind tests. Which of the following choices correctly describes a double-blind test?
    a) The researchers apply two layers of blindfolds to the study's participants so they don't know if they are receiving the drug or a placebo.
    b) The researchers do not know who receives the drug or the placebo but the participants know and tell them later.
    c) Neither the researchers nor the study's participants know who is receiving the drug and who is receiving the placebo.
    d) The researchers know who is receiving the drug and who is receiving the placebo but do not know what the supposed effects of the drug should be.
    e) None of the above descriptions is correct.

8. Which of the following is a limitation on scientific research?
    a) It is difficult for scientists to fully understand their own motivations and subjective biases given the complexity of human behavior.
    b) Scientists have to reduce complex phenomena to simple, testable hypotheses.
    c) Scientific researchers have ethical and legal responsibilities that can constrain their work.
    d) Scientific research cannot answer value-based problems.
    e) All of the above are limitations on scientific research.

Short Answer

1. You are studying the effects of artificial sweetener on mice. Four groups of mice consume different amounts of sweetener in their food. How much sweetener would the control group receive? Explain your answer.

2. You are testing treatments for cancer patients and find that 75% of patients respond very well to a particular treatment while 25% show no improvement or decline in health after taking the experimental treatment. What should you do next?

3. It is notoriously difficult to perform unbiased studies when human subjects are used. When governmental agencies (like the FDA) are trying to determine the validity of scientific claims, they very closely examine the method in which the experimental studies were done. Name several things that would be important in evaluating how the studies were performed.

4. There are many nutritional supplements on the market that allege that they can cause weight loss without dieting or exercise. Suppose you read a claim that a particular supplement causes weight loss and you are given the following information:

    * People were weighed at the beginning of the study.
    * People were asked to take two pills per day.
    * People were weighed at the end of the study.
    * People who took the supplement for four weeks seemed to have lost some weight at the end of the study.
    * It was concluded that the supplement is helpful for weight loss.

    This study has some obvious holes in it. Explain five things that could be done to this study to improve it. Do not assume any information other that what has been provided.

5. Explain the importance of placebos in human clinical trials. Are placebos a form of treatment? Explain your answer.

6. Suppose that data are collected that appear to link to variables. The data suggest that as hours of sleep per night decline, reaction times to various stimuli also decrease. Is this an example of a positive or negative correlation? Explain your answer.

7. Your roommate exhibits a particular ritual before each exam she takes because she is convinced that this ritual ensures her success on exams. She eats the same meal the night before, goes to bed at exactly the same time, listens to the same song the morning of the exam, and always wears the same sweatshirt to the exam. Design an experiment that will help to determine if her ritual does in fact enhance her exam performance.

8. It is not uncommon to see statistics manipulated in the media and in marketing in order to convince people that a certain treatment has a more pronounced effect than it really does. An example is a weight loss supplement advertised that has caused 514% more weight loss than a competing product. What information do you need to know in order to put this statistic in context?

Chapter 2
CHEMISTRY: RAW MATERIALS AND FUEL FOR OUR BODIES

Learning Objectives

- Relate major principles of chemistry to the study of biology
- Understand atomic structure
- Compare and contrast the major chemical bonds involved in important molecules and compounds
- Identify the major macromolecules that make up living organisms
- Explain the importance of the function of each of the four major macromolecules
- Describe the different categories of carbohydrates and their uses
- Compare and contrast different categories of lipids and their structures
- Understand the importance of shape in the functioning of a protein.
- Compare and contrast DNA and RNA

Chapter Outline

I. Atoms Form Molecules through Bonding

- Understanding chemistry is essential to the study of biology. Learning about chemistry allows us to understand the necessary building blocks of important molecules and the chemical reactions that help power a cell.

- Long regarded as precious, gold has always been evaluated for its beauty and value. However, when evaluating gold for its chemical makeup, gold is an example of an _____, or a substance that can not be broken down further into any other substance.

A. Everything Is Made of Atoms

- Describe the relationship between an **atom** and an **element**.

- The individual particles, or pieces, that form the structure of an atom have unique characteristics. Complete the following chart to highlight the particles that form the structure of an atom.

Particle type	Charge

  - Which two particles make up the **nucleus** of the atom?

- Define the **atomic mass** of an atom.

  - Do electrons contribute to the overall mass of the atom? Explain.

- Carbon has an **atomic number** of six. This number refers to the number of _____ carbon has. Will any other atom of a different element have the same number? Explain.

- Carbon-13 and carbon-14 are examples of _____ of carbon. These atoms have an unequal number of neutrons and protons.

- **Radioactive** isotopes are unstable and spontaneously release high energy particles. A radioactive atom may release protons or neutrons in an attempt to achieve stability.
  - Explain how an atom would change by losing neutrons.

  - Conversely, explain how an atom would change by losing protons.

- All the known elements are organized together on the **periodic table** (available in the appendix of your book). When examining the human body we find we are composed of 25 different elements. Of those 25, only 4 elements make up approximately 96% of us! List the "Big 4" here.

    1.

    2.

    3.

    4.

B. Atoms' Electron Shells

- _____ determines how and if an atom will bond with other atoms

- Because the aforementioned particle carries a negative charge, there is a limit to the number that move around the nucleus of the atom together, or in the same electron shell.

    o The atom's first electron shell, which is innermost and closest to the nucleus, can hold _____ electrons

    o The second and subsequent shells can hold more, or up to _____ electrons.

- Draw the appropriate number of electron shells and the appropriate number of electrons in each shell for Carbon. The nucleus of the atom is below. (Hint: Carbon has six electrons)

Ⓒ

- Explain why the atom is stable when the outermost electron shell is full.

    o What happens if the outermost shell is empty?

- In your own words, briefly describe what an atom has to do to become an **ion**.

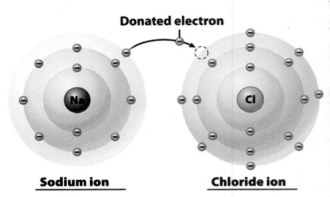

Donated electron

Sodium ion   Chloride ion

C. Chemical Bonding

- A **molecule** is a group of _____ held together by _____.

- What determines the type of bond that will form between reactive atoms?

- Types of Bonds – Describe how each bond holds atoms or groups of atoms together and list an example.

    1. Covalent Bond

    2. Ionic Bond

    3. Hydrogen Bond

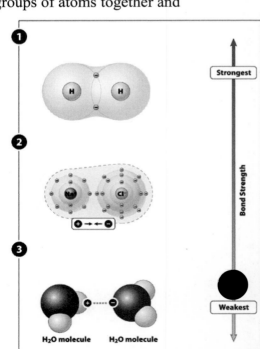

II. Water Has Features That Enable It To Support All Life.

A. Hydrogen Bonds Make Water Cohesive.

- Explain, using the term *polar molecules*, how organisms such as the fishing spider can take advantage of surface tension and walk across water.

B. Water Has Unusual Properties That Make It Critical To Life.

- List the four important properties of water and then explain how the feature is essential for living organisms.

1.

2.

3.

4.

C. Living Systems Are Highly Sensitive To Acidic And Basic Conditions.

- A water molecule can be broken apart releasing _____ and _____ ions.

- These ions individually can affect the pH of a particular solution. The **pH** scale measures

  _____.

- A fluid with a pH between 0 and 6.9 would be considered _____ and has more H^+ or OH^- ions?

- A fluid with a pH between 7.1 and 14 would be considered _____ and has more H^+ or OH^- ions?

- Give your own example of a food or household item that would register as an acid on the pH scale. Give an example of an item that would register as basic.

- Explain what happens to a solution when you add a buffer.

III. Carbohydrates are Fuel for Living Machines
A. Carbohydrates Include Macromolecules That Function as Fuel.

- A **macromolecule** can be simply defined as:

- The four major types of macromolecules that are necessary for our cells include
  1.

  2.

  3.

  4. **Carbohydrates**
     a. Carbohydrates are made up of the three following elements
         i.
        ii.
        iii.
     b. All carbohydrates are found with a similar composition of an H_2O unit for every carbon. Give the chemical formula for glucose: _____

     c. Monosaccharides are also called _____. Give an example of a monosaccharide you are familiar with:

    d. What is the primary function of carbohydrates in our cells?

        i. Explain why carbohydrates are so well suited for this function.

B. Glucose Provides Energy for The Body's Cells.

- The glucose from fruit, bread, potatoes and other carbohydrate sources ends up in the bloodstream. There are then several potential end results for the glucose circulating in the blood. Explain the three potential fates:

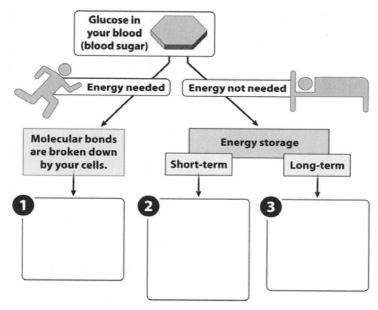

- What is glycogen's role in "carbo-loading"?

- What does "water weight" have to do with glycogen?

C. Complex Carbohydrates Are Time-released Packets of Energy.

- Complex carbohydrates are a source of fuel containing more than one sugar building block. These include:

    1. _____ which are two monosaccharides linked together and,

    2. _____ which are many monosaccharides (even thousands) linked together.

- **Disaccharides**
    - Give an example of a disaccharide and explain how it is utilized for fuel.

- **Polysaccharides**
    - What is the function of **starch** in plants?

    - What is the function of glycogen in humans?

D. Not All Carbohydrates Are Digestible.

- There are unique polysaccharides that are indigestible by humans but can serve as structural material for various organisms. Give an example of where each of the following are found.
    - **Chitin**:

    - **Cellulose**:

        - Why is cellulose an important part of the human diet even though humans can not digest it?

IV. Lipids Store Energy for a Rainy Day

A. Lipids Are Macromolecules with Several Functions, Including Energy Storage.

- **Lipids** are a large, diverse group of macromolecules with important functions. Like other macromolecules the structure of lipids includes the elements C, O, and H, but lipids have a higher ratio of C-H bonds. How does this affect lipids?

- Some important properties of lipids include how they behave in water. Define:
    o **Hydrophobic**

    o **Hydrophillic**

- List the three types of lipids and their function.

B. Fats Are Tasty Molecules Too Plentiful in Our Diets.

- The fats we are concerned about in the foods we eat are also called
_____.

- List the parts of the structure of a triglyceride and label these parts on the symbol below.

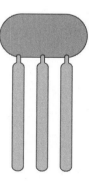

- Explain the structural difference between a saturated fat and an unsaturated fat.

    o   What is a favorite food of yours that is high in saturated fats?

    o   What is a favorite food of yours that is high in unsaturated fats?

- How is trans fat different from both saturated and unsaturated fats?

- Olestra is a fat substitute. It is found in many food products, such as potato chips, that strive to reduce the fat content in order to make the product more appealing to consumers.
    o   What is the chemical structure of Olestra?

    o   Are there consequences associated with a diet high in Olestra? Explain.

C. Cholesterol And Phospholipids Are Used to Build Sex Hormones And Membranes.

- What is the basic structure of all sterols?

    o   List two important examples.

- Draw the basic structure of a phospholipid.

    o   This category of lipids plays an important role in the cell's
    _____.

- Explain how **waxes** differ structurally from fats. Why are they sometimes found on the surface of plants and insects?

V. Proteins Are Versatile Macromolecules That Serve as Building Blocks.

- Proteins are the most diverse groups of macromolecules and function to regulate cellular activity, provide support, and assist in chemical reactions.

A. Proteins Are Bodybuilding Macromolecules.

- Proteins are similar to carbohydrates and lipids as they also contain C, O, and H, but they can be distinguished from these other important macromolecules because they also contain _____.

- Despite the large variety of proteins found, they are all composed of the same building blocks. By connecting different _____ _____ together, the result is a unique protein. There are _____ different amino acids available, just as there are 26 letters in our alphabet to select from when writing different words.

- Structure of an **amino acid**
  - o Identify the three main groups that form the structure of the amino acid

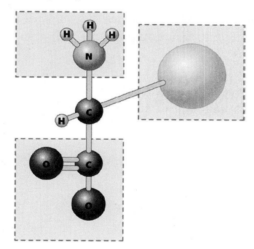

  - o What makes one amino acid different from another?

B. Proteins Are an Essential Dietary Component.
- Our cells require amino acids for various cellular structures and activities. Cells can make some of the amino acids that are required, but not all of them. Other necessary amino acids are obtained through the diet. The amino acids we must be certain to include in our diet are called _____.

- Briefly describe the difference between a complete protein and incomplete protein.

C. A Protein's Function Is Influenced by Its Three-dimensional Shape.

- Cells will build protein by linking amino acids together by a _____ bond.

- The order of the amino acids as well as the overall structure the protein takes on impacts how that protein functions.
    - o   The possible shapes, or structure, of proteins are:
        1.
        2.
        3.
        4.
    - o   If at any time, something alters the shape of the protein, the protein will no longer be able to function properly. That, in turn, can affect cellular functioning.

- When a protein is exposed to an extreme environment, such high temperature or changes in pH, what is the term that is used to describe what has occurred?

    _____

    - o   Explain why we call the clear or opaque protein of an egg, an egg "white".

D. Enzymes

- One important class of proteins is **enzymes**. Briefly *explain* what an enzyme's job is and if they can perform their job more than once.

- Only specific **substrates** can fit in an enzyme's _____ site to catalyze a reaction.

- **Activation energy** is required to jump start a reaction. What is the relationship between enzymes and a reaction's activation energy? Include four specific examples below.

  1.

  2.

  3.

  4.

D. **Enzymes Regulate Reactions in Several Ways**

- We know enzymes are needed for cellular reactions, but there are also other important environmental factors to consider. Environmental factors can influence the rate at which an enzyme catalyzes a reaction.

  Identify these factors and give a brief description of their impact on a reaction.
  1.

  2.

  3.

  4. Presence of inhibitors or activators. Differentiate between competitive and noncompetitive inhibitors

VI. Nucleic Acids Store The Information on How to Build and Run A Body.

A. Nucleic Acids Are Macromolecules That Store Information

- The fourth major category of important macromolecules is **nucleic acids**. Two types of nucleic acids are:
  1.

  2.

- The building blocks for nucleic acids are _____.

- A nucleotide is made up of three parts
  1. A phosphate group (PO4)
  2. A sugar
  3. A nitrogen containing molecule called a _____.

      o Each nucleotide can contain one of _____ different bases. These bases include:

        1.

        2.

        3.

        4.

B. DNA Holds The Genetic Information to Build an Organism

- Briefly explain the job of DNA in producing proteins.

- Many nucleotides linked together create a strand of DNA. One strand of DNA is bound to a second strand of DNA by _____ bonds. These two strands then spiral around each other forming the structure of DNA known as a

  _____.

- It is often easier to draw DNA as if it were a ladder. The vertical uprights of the ladder would include the _____ and _____. The rungs of the ladder, or where you would place your foot if you were climbing the ladder, would be the _____.

- If you were to examine the structure of DNA is great detail you would start to notice a pattern of which DNA bases were connected, or how one strand of DNA was bound to the second strand of DNA. Explain which bases always pair together.

C. RNA Is a Universal Translator, Reading DNA And Directing Protein Production.

- Briefly explain the job of RNA in producing proteins.

- There are three major differences between DNA and RNA. List the ways RNA is unique:
  1.

  2.

  3.

Testing and Applying Your Understanding

Multiple Choice

1. Which of the following statements about enzymes is NOT true?

a) Enzymes speed up chemical reactions.
b) Enzymes reduce the energy difference between reactants and products.
c) Enzymes are biological catalysts.
d) Enzymes often induce conformational changes in the substrates to which they bind.
e) Enzymes lower the activation energy of chemical reactions.

2. One of the four nucleotide bases in DNA is replaced by a different base in RNA. Which base is it, and what is it replaced by?

a) adenine, replaced by uracil
b) thymine, replaced by guanine
c) guanine, replaced by cytosine
d) thymine, replaced by uracil
e) cytosine, replaced by guanine

3. Which statement about phospholipids is FALSE?

a) They are hydrophilic at one end.
b) Because their phosphate groups repel each other, they are used as organisms' chief form of short term energy.
c) They are a major constituent of cell membranes.
d) They contain glycerol linked to fatty acids.
e) They are hydrophobic on one end.

4. Evaporation from the leaves of a tree will pull water up through the roots as an unbroken column throughout the entire height of the tree. This feat is possible because of which characteristic of water?

a) cohesion
b) kinetic energy
c) surface tension
d) absorption
e) vaporization

5. The thing that distinguishes one element, such as chlorine, from another such as neon, is:

a) the number of protons and neutrons in the nucleus.
b) the number of protons in the nucleus.
c) the number of neutrons in the nucleus.
d) the number of electrons.
e) the number of protons, neutrons, and electrons.

6. One important difference between covalent and ionic bonds is that:

a) in covalent bonds two atoms share electrons while in ionic bonds one atom gives one or more electrons to the other atom.
b) ionic bonds only occur among water-soluble elements.
c) in ionic bonds two atoms share electrons while in covalent bonds one atom gives one or more electrons to the other atom.
d) ionic bonds are much stronger than covalent bonds.
e) in ionic bonds both protons and electrons can be shared while in covalent bonds only electrons can be shared.

7. Which of the following is a polysaccharide?

a) cellulose, the primary component of plant cell walls
b) glucose, the chief cellular energy source
c) insulin, the chief blood sugar regulator
d) fructose, one of the most important blood sugars
e) All of the above are polysaccharides.

8. Saturated fatty acids have _____ than unsaturated fatty acids, which is why they exist as a _____ at room temperature.

a) fewer hydrogen atoms; solid
b) more carbon atoms; solid
c) more double bonds; liquid
d) fewer double bonds; solid
e) more glycerol molecules; liquid

9. Proteins are an essential component of a healthy diet for humans (and other animals). Their most common purpose is to serve as:

a) inorganic precursors for enzyme construction.
b) fuel for running the body.
c) raw material for growth.
d) long-term energy storage.
e) organic precursors for membrane construction.

Short Answer

1. Using carbon as an example, identify the atomic mass and atomic number of the element. If this was a neutral atom, how many electrons would it contain? Explain your answer.

2. Explain the difference between a stable atom and an unstable atom.

3. Chlorine atoms have an atomic number of 17. Assuming there are two neutral atoms present, draw the electron shells of two chlorine atoms and predict the type of bond that will form between them. Explain your answer.

4. Explain the chemical properties that allow water to be a good solvent.

5. Your roommate is experiencing heartburn and asks you what she should take for some relief. Explain, chemically speaking, how an antacid will help eliminate her discomfort.

6. We know once glucose arrives in the cell it can be used to fuel cellular activities. Explain why, then, athletes consume meals high in starches prior to an event versus meals high in simple sugars or fibers.

7. When examining the packages of some snack foods that are made with Olestra you note that certain vitamins are included in the ingredient list but with the notation that they are dietarily insignificant. Why would the manufacturer be including vitamins in the ingredients of the snack food containing Olestra?

8. There are several major categories of hormones in the body that vary based on their monomers units, or building blocks. A researcher investigating the functionality of hormones heats estrogen, testosterone and insulin to a very high temperature. After heating the hormones, the researcher tests to see if they still work properly. She finds that estrogen and testosterone still function but insulin does not. Based on the chemical nature of these molecules, explain why insulin no longer functions.

Chapter 3
CELLS—THE SMALLEST PART OF YOU

Learning Objectives

- Explain the two principles of the cell theory
- Compare and contrast prokaryotic and eukaryotic cell structure
- Discuss the theories of endosymbiosis and invagination as a means to explain the presence of organelles in eukaryotic cells
- Describe the overall structure of the plasma membrane as well as the structure of individual phospholipids
- Describe the following passive transport methods: diffusion, facilitated diffusion, osmosis
- Describe what happens to cells in isotonic, hypertonic, and hypotonic solutions
- Contrast the various types of active transport methods
- Explain the various structures that allow for connection and communication between cells
- Describe the major landmarks of eukaryotic cells
- Explain the structural differences between animal and plant cells

Chapter Outline

I. What is a cell?

- Cells are the smallest independently functioning living units. Each cell can perform all basic functions of life including reproducing itself.

- The term "cell" was first used in the 1600s by _____ after viewing cork under a microscope.

- Cells are three-dimensional structures where chemical reactions occur. List some examples of the sorts of reactions that occur in cells.

- While a few types of cells are visible, most are very small. Why is it that most cells are so very tiny?

A. The Cell Theory

- Indicate the two major principles of the **cell theory**:

    1.

    2.

B. Prokaryotic Cells

- There are two basic types of cells. These are _____ and _____. Only one of them houses genetic material in a central control center called the _____. Circle the name of the cell type that contains this structure.

- Prokaryotes have existed for about _____ billion years, as compared to eukaryotes, which have existed for about _____ billion years.

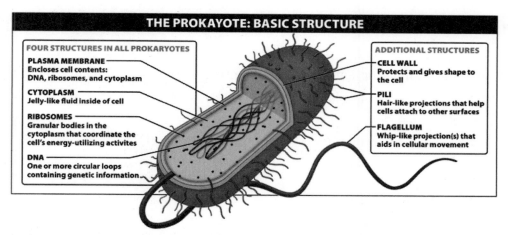

- Prokaryotic cells all contain four basic features as seen in Figure 3-3. Describe and indicate the role of each.

    1. **Plasma membrane:**

    2. **Cytoplasm:**

    3. **Ribosomes:**

    4. **DNA:**

- There are additional structures that some prokaryotic cells possess. Describe at least two of these optional structures.

- List the two major groups of prokaryotes:

    1.

    2.

- If you were to view an unknown cell type under the microscope, what features would be indicative of a prokaryotic cell?

- Can a multicellular organism be made of prokaryotic cells? Explain why or why not.

C. Eukaryotic Cells—The Basics

- Besides having a nucleus, eukaryotic cells contain a variety of membrane-bound **organelles** including a nucleus. What benefits do organelles confer for eukaryotic cells?

- As compared to a prokaryotic cell, eukaryotic cells are about _____ times larger in size.

- Are eukaryotes always multicellular? Explain why or why not and provide examples.

- List the features that differentiate animal cells from plant cells.

- The theory of **endosymbiosis** explains the presence of certain organelles. The two organelles in particular that developed through endosymbiosis are the _____ and the _____.
  Endosymbiotic theory is supported by several lines of evidence. These include

  1.

  2.

  3.

  4.

- Another theory to explain the presence of certain organelles in eukaryotic cells involves the **invagination** of membranes. What does invagination mean?

II. Cell Membranes Are Gatekeepers

- All cells contain a dual-layered **plasma membrane** that separates the interior of the cell from the external environment. This membrane has the essential functions of holding the contents of a cell in place and regulating passage into and out of the cell.

- List several examples of items that might need to cross the plasma membrane of a cell.

A. General Structure of the Plasma Membrane

- The basic structure of the plasma membrane consists of a **phospholipid bilayer** with fluidity. Each phospholipid is a dual-natured molecule composed of the following:

- o   The head is a **glycerol** molecule that contains a charge, making it
  _____ such that it is **hydrophilic** and interacts well with
  water.

- o   Two hydrocarbon tails that lack electrical charges, making them
  _____ and **hydrophobic**.

- The phospholipids orient themselves so that the heads are facing the aqueous
  environment inside and outside the cell while the tails are tucked together in the
  interior of the membrane.

B. The Role of Proteins in the Plasma Membrane

- In addition to the phospholipid bilayer structure of the plasma membrane, there
  are a variety of protein types found within the membrane. **Transmembrane
  proteins** span the entire membrane because their tertiary structure has
  hydrophobic and hydrophilic regions, while **surface proteins** reside on the
  exterior surface of the membrane due to their hydrophilic structure.

- How does the plasma membrane stay attached to the rest of the cell?

- Explain the role of each of the four major categories of proteins (seen in Figure 3-
  11) found in the plasma membrane.

  1. **Receptor proteins:**

  2. **Recognition proteins:**

  3. **Transport proteins:**

4. Enzymatic proteins:

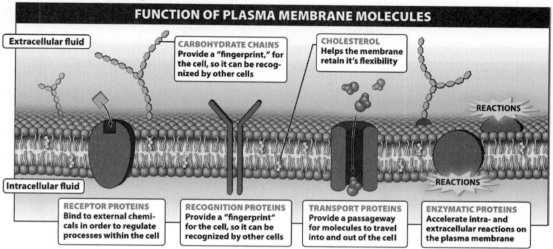

FUNCTION OF PLASMA MEMBRANE MOLECULES

Extracellular fluid

CARBOHYDRATE CHAINS
Provide a "fingerprint," for the cell, so it can be recognized by other cells

CHOLESTEROL
Helps the membrane retain it's flexibility

REACTIONS

Intracellular fluid

REACTIONS

RECEPTOR PROTEINS	RECOGNITION PROTEINS	TRANSPORT PROTEINS	ENZYMATIC PROTEINS
Bind to external chemicals in order to regulate processes within the cell	Provide a "fingerprint" for the cell, so it can be recognized by other cells	Provide a passageway for molecules to travel into and out of the cell	Accelerate intra- and extracellular reactions on the plasma membrane

- Proteins and phospholipid heads within the plasma membrane can be modified with carbohydrate chains. These modified molecules serve as identifying markers for the cell, which can be recognized by the immune system. This can be beneficial in terms of fighting foreign cells that cause disease but bad in terms of situations, such as transplants, where we would like foreign cells to be able to exist in the body.

- **Cholesterol** can also be present in the plasma membrane. The role of cholesterol is to:

- The plasma membrane is often referred to as a fluid mosaic. Explain what this term means.

C. Membrane Malfunctions

- Cystic fibrosis is a genetic disease that involves a faulty transmembrane protein.

    o   What is the role of this protein under normal circumstances?

    o   How is the protein altered in a person with cystic fibrosis?

    o   How does this lead to the symptoms of the disease?

- Explain how the beta-blocker category of drugs works to reduce blood pressure and anxiety.

D. Membrane "Fingerprints"

- What comprises the "fingerprints" on the surface of your cells?

- How do membrane "fingerprints" cause problems with organ transplants? What do we try to do to get around this rejection?

- Explain the role of membrane cluster of differentiation (CD) markers, specifically CD4, that allows for the infection of cells by the HIV virus.

- What are the typical modes of transmission of the HIV virus?

III. Molecules Move across Membranes in Several Ways

- The movement of molecules into and out of the cell can occur in one of two ways. When energy is required, the movement is termed _____ transport. When energy is not required, the movement is termed _____ transport.

A. Passive Transport

- **Diffusion** is the spontaneous movement of a substance called a _____ across the plasma membrane from the side of the membrane that has a higher concentration of the substance to the side of the membrane that has a lower concentration of the substance. In this case, the substance is moving down its _____.

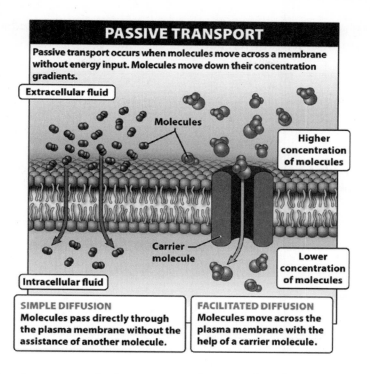

PASSIVE TRANSPORT

Passive transport occurs when molecules move across a membrane without energy input. Molecules move down their concentration gradients.

Extracellular fluid

Molecules

Higher concentration of molecules

Carrier molecule

Lower concentration of molecules

Intracellular fluid

SIMPLE DIFFUSION
Molecules pass directly through the plasma membrane without the assistance of another molecule.

FACILITATED DIFFUSION
Molecules move across the plasma membrane with the help of a carrier molecule.

- When molecules are passively transported across the plasma membrane through a carrier protein, this is termed _____ diffusion. Both forms of diffusion can be seen in Figure 3-17.

- A specialized form of diffusion involves the movement of water across the plasma membrane from an area of high water concentration to an area of low water concentration. This form of transport is referred to as **osmosis**.

  o While osmosis involves only the movement of water, it is influenced by the concentration of other molecules inside or outside the cell that are unable to move across the membrane.

  o The term _____ refers to the relationship between the concentrations of solutes inside the cell and the solutes outside the cell.

  o Give two reasons that a particular molecule may be unable to cross the plasma membrane.

    1.
    2.

  o In terms of osmosis, water will move into a cell that is placed in a _____ solution, while water will exit a cell that is placed in a _____ solution.

o When cells are placed in fresh water, what happens to them in terms of osmosis? What differences are seen between plant and animal cells?

o Provide a practical example of osmosis and explain how it works, using appropriate terminology.

B. Active Transport

- _____ requires energy in order to move large molecules or to move substances against their concentration gradients.

- _____ active transport relies directly on ATP to fuel molecular movement across the membrane, while _____ active transport couples the movement of one substance across the membrane moving down its concentration gradient while another substance moves against its concentration gradient.

- The proton pumps in your stomach are examples of active transport. How do these pumps work? Is this primary or secondary active transport?

- Molecules that are very large can only be transported in certain ways. Movement of large molecules into the cell is called _____, while the movement of large molecules out of the cell is called _____.

- Explain the role of vesicles in bulk transport.

- **Endocytosis** can occur in several ways. Describe each.

  o **Phagocytosis:**

o **Pinocytosis:**

o **Receptor-mediated endocytosis:**

- Provide an example of **exocytosis** in cells.

IV. Cells Are Connected and Communicate with Each Other

- There are several connections that occur between cells in order to connect the cells to each other and to facilitate communication between the cells. Fill in the following table concerning connection types between cells.

Type of connection	How it works	Example
Tight junction		
Desmosomes		
Gap junction		

V. Nine Important Landmarks Distinguish Eukaryotic Cells

- Eukaryotic cells have a variety of membrane-bound _____, which distinguish them from prokaryotic cells.

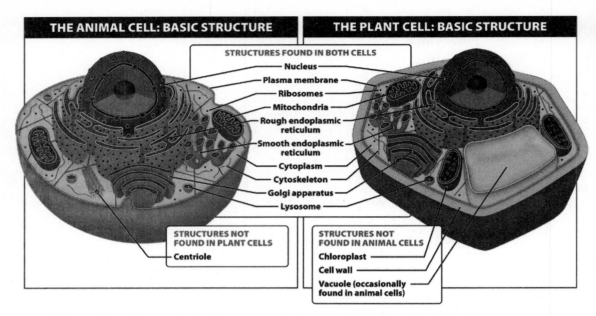

- Eukaryotic cells come in the two varieties seen in Figure 3-6. For the most part, they share the same features with only a few structural differences between the two types. Plant cells contain cell walls, chloroplasts, and central vacuoles, while animal cells do not. Animal cells contain centrioles which plant cells do not.

A. The Nucleus

- The **nucleus** houses the genetic material of a cell, directing the synthesis of proteins within the cell.

- The nucleus is kept separate from the rest of the cell by the _____ membrane. It is structured like the plasma membrane except it is composed of two bilayers of phospholipids. It contains pores to allow substances to move out of the nucleus to the cytoplasm.

- DNA is found within the nucleus. It associates with a variety of proteins to keep it organized and prevent breakage. In this form, the DNA is referred to as

  _____.

- The _____ is found within the nucleus where it produces the subunits needed to produce **ribosomes**.

B. The Cytoskeleton

- The cytoskeleton is made of several types of proteins that form an internal scaffolding for the cell. The three purposes of the cytoskeleton are as follows:

1.

2.

3.

- Describe the function of each of the three types of cytoskeletal elements:

    1. Microtubules:

    2. Intermediate filaments:

    3. Microfilaments:

- Some cells have additional cytoskeletal elements.

    o **Cilia** are short projections that occur on the surface of cells. What purposes do cilia serve?

    o **Flagella** are longer projections extending from the surface of cells. What function do flagella provide for a cell?

C. Mitochondria

- **Mitochondria** serve a critical role in eukaryotic cells by producing energy from carbohydrates, fats, or proteins.

- When the mitochondria produce energy, three products are produced. Those include ATP, _____, and _____. In order to produce ATP, the mitochondria must have a constant source of oxygen.

Different types of cells have different numbers of mitochondria. Which types of cells in the human body would have a lot of mitochondria based on their energy demands? Which types of cells might have less mitochondria?

- The mitochondria have a double membrane structure where one membrane is essentially wadded up inside another membrane. The interior of the organelle is referred to as the _____. Describe the benefits that are afforded by this double-membraned structure.

- Endosymbiotic theory can explain why mitochondria have their own DNA. While we are used to thinking about our DNA being equally contributed by both parents, this is not the case with mitochondrial DNA. Which parent do you receive your mitochondrial DNA from? Why is this the case?

D. Lysosomes

- Lysosomes act as the garbage disposal of the cell. The interior of these organelles contains a variety of digestive enzymes in a very acidic solution. What sorts of substances might the lysosomes dispose of?

- Why is it absolutely essential that the lysosomes need to be walled off from the rest of the cell by a membrane?

- While the term lysosome is used to be restricted to animal cells, we now know that plant cells also have lysosomes..

E. The Endoplasmic Reticulum

- The **endoplasmic reticulum** (ER) is part of the endomembrane system—a series of organelles involved in synthesis and modification of molecules within cells. The ER is large and is situated near the nucleus.

- The **rough endoplasmic reticulum** appears "rough" due to the presence of _____ on its surface.

  o What is the primary activity that occurs in the rough ER?

- The **smooth endoplasmic reticulum** deals in the synthesis of
_____.

  o The smooth ER in liver cells has a very specialized role, which is:

F. The Golgi Apparatus

- The **Golgi apparatus** is another part of the endomembrane system. The contents of the rough ER and smooth ER are moved to the Golgi apparatus via transport _____.

- What happens to molecules in the Golgi apparatus?

- What are the potential fates of items that exit the Golgi apparatus?

G. Structures Unique to Plant Cells

- Plants have some unique needs that require adaptations and structures not seen in animal cells. These structures include the cell wall, the central vacuole, and chloroplasts.

- The **cell wall** is made of _____, which is a carbohydrate. This structure provides several types of protection to plant cells. What sorts of things does the plant cell wall provide protection against?

- The **central vacuole** is a structure that takes up much of the volume in the plant cell. It has five specific functions. Explain each.

    1. Nutrient storage:

    2. Waste management:

    3. Predator deterrence:

    4. Sexual reproduction:

    5. Physical support:

- The **chloroplasts** in cells are responsible for photosynthesis. Like mitochondria, chloroplasts have a double membrane that can be explained by the theory of endosymbiosis.

    o   The inner compartment of chloroplasts is filled with a fluid called _____ where the DNA is located. This is also the site of sugar production during photosynthesis.

    o   Inside each chloroplast are stacks of small discs called _____. These structures are responsible for the collection of solar energy for photosynthesis.

Testing and Applying Your Understanding

Multiple Choice (For more multiple choice questions, visit www.prep-u.com.)

1. A scientist tries to build a eukaryotic cell. She remembers to include most of the organelles, but forgets one. Her newly created cell cannot synthesize the enzymes needed to detoxify drugs and poisons. Which organelle is missing?
    a) the mitochondria
    b) the nucleus
    c) the Golgi apparatus
    d) the cytoskeleton
    e) the smooth endoplasmic reticulum

2. Which of the following cannot occur?
    a) You and your paternal aunt having identical mitochondrial DNA.
    b) Two cousins having identical mitochondrial DNA.
    c) You and your full-sibling brother having different nuclear DNA.
    d) You and your great-grandmother having identical mitochondrial DNA.
    e) None of the above is correct; all of these are possible.

3. Given that a cell's structure reflects its function, what would you predict that the function of a cell with a large Golgi apparatus would be?
    a) attachment to bone tissue
    b) movement
    c) secretion of digestive enzymes
    d) rapid replication of genetic material and coordination of cell division
    e) transport of chemical signals

4. In eukaryotic cells, vesicles connect which of the two major organelle compartments?
    a) the rough endoplasmic reticulum and the nucleus
    b) the rough endoplasmic reticulum and the centrioles
    c) the rough endoplasmic and the ribosomes
    d) the smooth endoplasmic reticulum and the lysosomes
    e) the rough endoplasmic reticulum and the smooth endoplasmic reticulum

5. What does it mean to be "alive"? Scientists have outlined some basic characteristics that must be present in order for an object to be considered alive. Which of the following choices is NOT one of these basic characteristics?
    a) The object must use and metabolize some form of energy.
    b) The object uses oxygen in some form.
    c) There is some level of organization present in the object's components.
    d) The object responds in some manner to stimuli and adapts to its environment.
    e) Some form of reproduction, growth, and development must occur with the object.

6. According to the theory of endosymbiosis, the origin of chloroplasts probably involved:
    a) the accumulation of free oxygen in ocean waters.
    b) the formation of colonies of cyanobacteria.
    c) the formation of cell walls around the photosynthetic pigments.
    d) engulfing of small photosynthetic prokaryotes by larger cell.
    e) All of the above were involved in the origin of chloroplasts.

7. Which of the following best summarizes the differences between osmosis and diffusion?
    a) Diffusion deals only with water.
    b) Osmosis deals only with acidic liquids.
    c) Diffusion deals only with alkaline liquids.
    d) Osmosis deals only with alkaline liquids.
    e) Osmosis deals only with water.

8. In Tay-Sachs disease, a genetic mutation causes a malfunction in an enzyme found in a certain organelle that leads to a backup of molecules and proteins, ultimately interfering with the entire functioning of the cell. Which organelles are primarily affected by this disease?
    a) the nuclei
    b) the lysosomes
    c) the mitochondria
    d) the peroxisomes
    e) the Golgi apparatus

9. You measure the concentration of a polar molecule inside and outside of a cell. You find that the concentration is high and gradually increasing inside the cell. You also measure the ATP concentration inside the cell and find that it is dropping. Your best hypothesis for the process that is occurring would be?
    a) facilitated diffusion
    b) passive transport
    c) active transport
    d) simple diffusion
    e) endocytosis

10. The passive transport of water across a membrane from a solution of lower solute concentration to a solution of higher solute concentration is best described as:
    a) passive transport
    b) simple diffusion
    c) facilitated diffusion
    d) osmosis
    e) active transport

Short Answer

1. As a human, you are composed of about 10 trillion cells. However, you carry about ten times that number of bacteria in and on your body. Based on what you know about cell structure, how is it that you can carry so many of these cells on your body?

2. You are observing an unknown cell type. It is very tiny but you are able to see it without a microscope. What sort of cell is this? Justify your answer.

3. You are viewing a picture of a cell taken from a very powerful microscope. The picture shows cytoplasm, ribosomes, and a cell membrane. Based on this information alone, can you determine which type of cell you are observing? Explain your answer.

4. A person who has had an organ transplant will typically need to take immuno-suppressing drugs for the rest of their life. This is to prevent the recipient's immune system from attacking the cells of the new organ. Without these drugs, the recipient's immune system will recognize certain structures in the donor cell's membranes as foreign. What are the specific membrane structures that will be recognized by the immune system?

5. Fill in the following table concerning osmosis.

Cell contains:	Solution outside the cell contains:	What type of solution is this?	Water will move in what direction?
10% solute 90% water	30% solute 70% water		
10% solute 90% water	5% solute 95% water		
5% solute 95% water	5% solute 95% water		

6. When baking with fruit, it is fairly common to prepare the fruit by cutting it and adding sugar. This has the effect of extracting the juices from the fruit. Based on what you know about osmosis, how can this be explained?

7. Neurons (cells within the nervous system) contain a high concentration of potassium ions as compared to their environment. Based on what you know about membrane transport, how could a neuron acquire even more potassium ions?

8. Pancreatitis is a disorder where pancreatic cells destroy themselves. This disorder is related to the rupture of a specific type of organelle. Considering what you know about organelle function, explain which organelle might be involved in pancreatitis.

9. A certain type of immune system cell makes antibodies (a type of defensive protein) that are secreted from the cell. It is possible to trace the path of these proteins through the cell by labeling them with radioactivity. Indicate the pathway these antibodies would take, starting from where they are made to their exit from the cell. Make sure to indicate each structure or organelle along the way.

10. At a party last weekend, your roommate had a few "adult beverages." A mutual friend makes the comment, "John couldn't drink like that last semester—he sure has developed a tolerance." On a cellular level, what has caused John to develop this newfound tolerance to alcohol?

Chapter 4
ENERGY—FROM THE SUN TO YOU
IN JUST TWO STEPS

Learning Objectives

- Understand that energy from the sun fuels all life on earth
- Describe the various types of energy and how energy can be stored in different molecules
- Explain the laws of thermodynamics and how they relate to living organisms
- Describe the locations where the reactions of photosynthesis and cellular respiration occur in cells
- Differentiate between the "photo" and "synthesis" reactions of photosynthesis
- Explain how oxygen gas is produced in photosynthesis
- Describe how CO_2 is incorporated into sugars during the "synthesis" reactions of photosynthesis
- Contrast C3, C4, and CAM photosynthesis
- Understand which types of organisms perform photosynthesis and cellular respiration
- Describe each of the steps involved in cellular respiration and the sorts of fuels that can be used for this process
- Explain how the electron transport chain produces ATP and the role of oxygen in the electron transport chain
- Describe how organisms survive under anaerobic conditions by using alternate energy pathways
- Compare and contrast the starting and ending materials in photosynthesis and cellular respiration

Chapter Outline

I. Energy Flows from the Sun and Through All Life on Earth

- Differentiate between a **biofuel** and a **fossil fuel** by explaining how each one is produced.

- Why are biofuels considered renewable sources of energy?

- The use of fossil fuels and biofuels both have their drawbacks. Indicate at least one drawback for the use of each type of fuel.

- In biofuels, fossil fuels, and food fuels, energy from the _____ serves as the source of energy stored in the chemical bonds of the fuel.

- When hydrocarbons, long chains of carbon and hydrogen, are broken down, energy is released. In addition to energy being released, two other byproducts are formed. These two items are:

- Two processes are required to harness the energy of the sun and bring it to you. Indicate those steps along with a generalized description of each:

    1.

    2.

A. Kinetic and Potential Energy

- In terms of **energy**, work is defined as:

- How do **kinetic energy** and **potential energy** differ?

- Provide two of your own examples of kinetic energy and potential energy:

    1. Kinetic energy:

    2. Potential energy:

- The chemical energy stored in chemical bonds is a specific type of _____ energy.

B. Energy Conversions Are not 100% Efficient

- Only about _____% of energy released by the sun is captured by living organisms, specifically plants. What happens to the remainder of solar energy?

- _____ is the study of transformation of energy from one form to another.

- The **first law of thermodynamics** states:

- The **second law of thermodynamics** states:

C. Adenosine Triphosphate (ATP)

- Energy from the sun and food cannot be directly used to fuel the chemical reactions in cells. Instead, this energy must be captured in the bonds of _____, which is a universal source of energy for all living things.

- **Adenosine triphosphate** is composed of three component molecules:

       1.

       2.

       3.

- Why is it that the bonds between the PO_4 groups in ATP contain a lot of energy?

- When an ATP molecule releases its energy, the resulting products are _____ and _____. How are these products recycled back to ATP?

- Explain three situations in your own body where ATP would be utilized.

      1.

      2.

      3.

- ATP is considered to be a source of _____ energy. When ATP is converted to ADP, _____ energy is released.

II. Photosynthesis Uses Energy from Sunlight to Make Food

- Plants are well-known photosynthesizers, but other organisms are also capable of performing photosynthesis. Some examples of those other organisms are:

A. The Big Picture

- The big picture overview of photosynthesis is presented in Figure 4-11. The process of **photosynthesis** can be broken down to two sets of reactions.

  o The term "photo" means _____.
    Explain what happens in the photo segment:

  o The term "synthesis" means _____.
    Explain what happens in the synthesis segment:

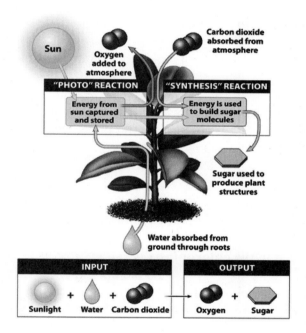

- Three inputs are needed for photosynthesis to occur. Explain where each of these inputs comes from and what it will be used for in photosynthesis:

  1. Light energy:

  2. Water:

  3. Carbon dioxide:

- There are two outputs produced during photosynthesis. Explain how each is created during photosynthesis and what will happen to each.

    1. Sugar:

    2. Oxygen:

- Describe the big picture significance of photosynthesis:

B. Photosynthesis and Chloroplasts

- How can you tell which parts of a plant are photosynthesizing?

- List several examples of plant tissues that would not perform photosynthesis. If these parts of the plant are not performing photosynthesis, what tasks might they perform for the plant?

- The organelle responsible for photosynthesis is the _____.

- The fluid located in the **chloroplasts** is called the _____. The membrane-enclosed network of discs floating in the fluid are called _____. Label these structures on Figure 4-13 and indicate in which locations the "photo" reactions and the "synthesis" reactions occur.

- In which part of the chloroplast would **chlorophyll** be found?

- In addition to making the photosynthesizing structures of a plant green, the role of chlorophyll is to:

C. Photosynthesis and Light Quality

- An energy packet which is a type of kinetic energy is referred to as a
  _____.

- The longer the wavelength, the _____ (more or less) energy the light carries.

- How is the electromagnetic spectrum organized?

- Explain how **chlorophyll *a*** and **chlorophyll *b*** are different from one another.

- If an object is reflecting the blue wavelength of light, what color will it appear?

- If an item absorbs all wavelengths in the visible spectrum, what color will it appear?

- Plants can only harvest certain wavelengths of solar energy. The wavelengths that plants can utilize correspond to the colors _____ and _____. Most of the light a plant is exposed to is reflected and cannot be used for photosynthesis. Different types of plants have evolved different requirements for the amount of light they require in order to function optimally.

- Chlorophyll production in deciduous trees halts in the fall. At this point _____ **pigments** can be seen in the leaves.

D. Photons and Electrons

- When an electron in chlorophyll absorbs a **photon**, the electron gains energy, which increases the _____ energy of the chlorophyll.

- The two possible fates of an excited electron are:

    1.

    2.

- In a chain of electron carrier molecules, would the first or the last electron carrier molecule have a greater affinity for electrons?

- Why is the passing of electrons a critical first step in photosynthesis?

- How can an increase in particles in the atmosphere impact the process of photosynthesis?

E. The "Photo" Reactions

- Light-catching pigments that capture solar energy in the chloroplasts are called _____. This harvesting of solar energy occurs in the _____ of the chloroplasts.

- Explain the difference between how chlorophyll *a* deals with excited electrons in the **photosystems** as compared to chlorophyll *b* and other pigments.

- Where does the **primary electron acceptor** in each photosystem collect its electrons from?

- As electrons are continually passed to primary electron acceptors, they must be replaced. How does this electron replacement work?

- How is oxygen gas produced in the "photo" reaction?

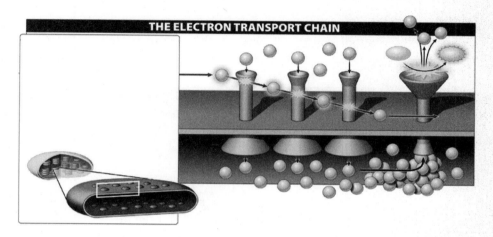

- Figure 4-19 shows the electron transport chain of the "photo" reactions of photosynthesis. Explain what is happening in this diagram.

    1.

    2.

    3.

- The first photosystem produces _____ while the second photosystem produces _____.

- What is the role of proton (H^+) pumps during electron transport chains?

- The terminal location of the electrons after they pass through both photosystems is:

F. The "Synthesis" Reactions

- The **Calvin cycle** occurs in the _____ of the chloroplasts.

- In order to carry out the Calvin cycle, the items needed from the "photo" reactions are _____and _____.

- Other than the items needed from the "photo" reactions, what is the additional item needed by plants in order to complete the Calvin cycle?

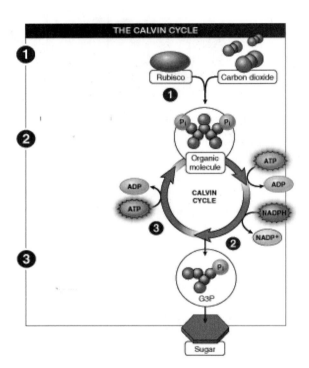

- The Calvin cycle has three major steps, as seen in Figure 4-22. Describe the details of each, making sure to notate any critical enzymes:

    1. Fixation

    2. Sugar creation

    3. Regeneration

- The Calvin cycle does not directly produce glucose for the plant. Instead, it produces a molecule called **G3P**. How do G3P molecules become sugar?

- Not all of the G3P molecules produced in the Calvin cycle are used to make sugar. What sorts of problems would occur if all of the G3P molecules were used to produce sugars for the plant?

- What happens to the sugars produced at the end of the "synthesis" reactions?

G. Alternate Pathways for Photosynthesis

- The major problem that plants in hot, dry environments face is _____ loss.

- Indicate the benefit a plant receives when its **stomata** are open.

- What is the drawback of plants leaving their stomata open?

- Explain how **C4 photosynthesis** is different from typical **C3 photosynthesis**.

    o   What is the trade-off in terms of energy for using this method?

- How is **CAM photosynthesis** unique?

    o   What disadvantages come from using this method of photosynthesis?

III. Cellular Respiration Converts Food Molecules to ATP, a Universal Source of Energy for Living Organisms

- What sorts of organisms perform **cellular respiration**?

- Why do organisms perform cellular respiration?

- Plants must perform photosynthesis as a precursor to cellular respiration. Why don't you, as an animal, have to do the same?

A. Glycolysis

- **Glycolysis** is the first step of cellular respiration for all organisms. The remaining steps of cellular respiration that follow glycolysis are the
  _____ and the
  _____.

- Glycolysis actually requires an ATP investment in order to begin the reactions even though the goal of cellular respiration is to produce ATP. What is this initial ATP investment used for and why it is necessary?

- The end result of glycolysis is two molecules of _____, which continue to the **Krebs cycle**, molecules of _____, which move to the **electron transport chain**, and _____ molecules, which can immediately be used for energy.

- Some organisms produce all the ATP they need just from glycolysis. Give a few examples of these sorts of organisms.

- When oxygen is unavailable, other organisms are unable to progress with additional steps in cellular respiration. Why does this present a problem for most organisms?

B. The Krebs Cycle

- While glycolysis occurs in the cytoplasm of cells, the remaining steps occur in the _____ of cells.

- Following glycolysis, three modifications to **pyruvate** must occur prior to the **Krebs cycle**. They are:

    1.

    2.

    3.

- After pyruvate is modified, the molecule _____ enters the Krebs cycle.

- Overall, the Krebs cycle involves a lot of rearrangements of molecules that allows for the transfer of high energy electrons. The molecules that are ready to accept these electrons are called _____ and _____.
  When these molecules actually pick up the electrons they are called
  _____ and _____.

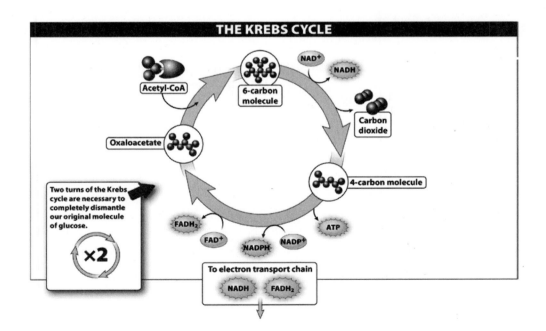

- Figure 4-32 shows the Krebs cycle. There are eight steps in this cycle but only three major outcomes. Explain the three major outcomes that are happening in the figure.

    1.

    2.

    3.

- Where does the CO_2 that you are exhaling right now come from? Be specific with your answer.

- This CO_2 you are exhaling will soon be utilized by plants. What will the plants ultimately do with the CO_2?

C. The Electron Transport Chain

- The **electron transport chain** serves as the site of the big ATP payoff in cellular respiration. The high-energy electrons needed for this step are delivered by _____ and _____ produced in glycolysis and the Krebs cycle.

- Explain the two features of the **mitochondrial** structure that contribute to its ability to produce energy in the electron transport chain.

- The electron transport chain is considered an aerobic process because it requires oxygen. This oxygen has a very specific job which is to:

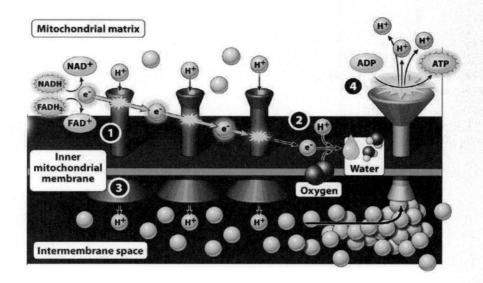

- Figure 4-34 shows the electron transport chain. Describe the four steps shown in this diagram.

    1.

    2.

    3.

    4.

- Once oxygen has performed its specific function, it combines with H$^+$ ions, ultimately producing _____.

- Approximately _____ ATP molecules can be produced through cellular respiration from a single molecule of glucose.

IV. There Are Alternative Pathways to Energy Acquisition

- Using glucose and oxygen to perform cellular respiration is not the only option cells have for cellular respiration!

A. Fermentation

- A lack of oxygen has major consequences for how cells perform cellular respiration. When oxygen is in short supply, a certain step of cellular respiration is impeded. That step is the _____.

- Under anaerobic conditions, there are two **fermentation** pathways cells can potentially use to survive. One of these pathways results in the production of _____ while the other pathway produces _____.

- All alcoholic beverages contain alcohol. Why do some taste different from others?

B. Alternate Fuels for Cellular Respiration

- The primary fuel for cellular respiration is _____ but cells can also use _____ or _____ as alternative fuels.

- Where do each of the two products of fat breakdown enter into the process of cellular respiration?

    o Glycerol

    o Fatty acids

- How can proteins be manipulated so that they can be used as an energy source in cellular respiration?

Testing and Applying Your Understanding

Multiple Choice (For more multiple choice questions, visit www.prep-u.com.)

1. Photosynthesizing plants rely on water:
    a) to replace electrons that are excited by light energy and passed from molecule to molecule down an electron transport chain.
    b) replenish oxygen molecules that are lost during photosynthesis.
    c) to provide the protons necessary to produce chlorophyll.
    d) to concentrate the beams of light hitting a leaf, focusing them on the reaction center.
    e) to serve as a high-energy electron carrier.

2. Although the reactions of the Calvin cycle do not depend directly on light, they do not usually occur at night. Why?
    a) It is usually too cold at night for these reactions to take place.
    b) Carbon dioxide concentrations decrease at night.
    c) The Calvin cycle depends on products of the light reactions that cannot occur at night.
    d) Plants must open their stomata to restore their water balance each night.
    e) At night, plants have a negative water balance with the soil, due to reduced evaporation.

3. If a thylakoid were punctured so that its interior was no longer separated from the stoma, which of the following processes would be most directly affected?
    a) the oxidation of NADPH
    b) the absorption of energy by chlorophyll
    c) the flow of electrons from photosystem I to photosystem II
    d) the synthesis of ATP
    e) the splitting of water

4. Cellular respiration is the process by which:
    a) energy from the chemical bonds of food molecules is captured by an organism.
    b) oxygen is produced during metabolic activity.
    c) ATP molecules are converted into water and sugar.
    d) light energy is converted into kinetic energy.
    e) oxygen is used to transport chemical energy throughout the body.

5. Glycolysis:
    a) is not performed in plants, which get their energy solely through photosynthesis.
    b) completely oxidizes glucose to carbon dioxide.
    c) is performed solely on the glucose ingested by the organism.
    d) is also referred to as the Krebs cycle.
    e) occurs in all cells.

6. Theodor Engelmann broadcast light that had been passed through a prism onto a mat of algae. This exposed different parts of the algae to different wavelengths of light. When he added aerobic bacteria to the system, he noted that the largest groups of bacteria aggregated in the areas of the algae illuminated by the blue and red light. What did Engelmann discover from this experiment?
   a) Bacteria aggregated in the area in which the most oxygen was being released.
   b) Bacteria aggregated in the area in which the most carbon dioxide was being released.
   c) Bacteria aggregated in the area with the highest concentration of photons.
   d) Bacteria aggregated in the area in which the most water was available.
   e) Bacteria had more chlorophyll *a* than the algae.

7. Which of the following statements most accurately depicts the relationship between the light reactions and the Calvin cycle in photosynthesis?
   a) The light reactions provide ATP and NADPH to the Calvin cycle and the Calvin cycle returns ADP, $NADP^+$, and a phosphate group to the light reactions.
   b) The light reactions provide ATP and NADPH to the carbon fixation step of the Calvin cycle, and the Calvin cycle provides water and electrons to the light reactions.
   c) The light reactions supply the Calvin cycle with CO_2 to produce sugars, and the Calvin cycle supplies the light reactions with sugars to produce ATP.
   d) The light reactions provide the Calvin cycle with oxygen for electron flow, and the Calvin cycle provides the light reactions with water to split.
   e) There is no relationship between the light reactions and the Calvin cycle.

8. During the Krebs cycle:
   a) the products of glycolysis are further broken down, generating additional ATP and the high-energy electron carrier NADH.
   b) the products of glycolysis are further broken down, generating additional ATP and the high-energy electron carrier NADPH.
   c) the products of glycolysis are converted into acetyl-CoA.
   d) high-energy electron carriers pass their energy to molecules of sugar, which store them as potential energy.
   e) cellular respiration can continue even in the absence of oxygen.

9. Which one of the following statements best represents the relationship between respiration and photosynthesis?
   a) Respiration occurs only in animals and photosynthesis occurs only in plants.
   b) Photosynthesis reverses the biochemical pathways of respiration.
   c) Respiration stores energy in complex organic molecules, while photosynthesis releases it.
   d) Photosynthesis stores energy in complex organic molecules, while respiration releases it.
   e) Photosynthesis occurs only in the day and respiration occurs only at night.

Short Answer

1. The amount of energy in food is measured in calories. The chemical bonds in food are broken to release energy that can be harvested for cells. Explain why it is that we commonly refer to this process as "burning" calories.

2. According to the second law of thermodynamics, the quality of energy changes over time. What sort of significance does this hold for living organisms?

3. Certain trees stop producing chlorophyll in the fall and eventually their leaves fall off so that they can no longer perform photosynthesis. How do they survive the winter?

4. A plant that keeps its stomata closed most of the time is going to be limited in a certain item needed for photosynthesis. What is this item and what reaction will be affected by a lack of it?

5. NAD^+ and FAD are created from certain B vitamins that you obtain in your diet. While NAD^+ and FAD are needed in abundance for your daily cellular respiration needs, the amount of the precursor vitamins you require is not very large at all. Why would this be?

6. Carbon monoxide poisoning often has fatal results. When an individual is exposed to large amounts of carbon monoxide, it essentially prevents oxygen from being delivered to electron transport chains. Based on your understanding of the electron transport chain, what specific problems would this cause? Why would the individual ultimately die?

7. Some of your friends are attempting to make their own wine. They start with grapes (a sugar source) and add yeast. However, when they check on their wine sometime later, they find that no ethanol is present. What might have prevented the yeast cells from producing the ethanol?

8. A plant is exposed to water labeled with a radioactive isotope of oxygen that can be traced as it moves through the plant. The plant is then exposed to sunlight. Once photosynthesis occurs, where will the radioactive oxygen from the water end up?

9. Plants are particular about the type of light they can use for photosynthesis. If you were to use a filter on a light so that the plant only received green wavelengths of light, what would you expect to happen to the plant and why?

10. Both photosynthesis and cellular respiration rely on electron carrier molecules to provide electrons to ultimately help produce ATP. In photosynthesis the source of electrons is from _____ and in cellular respiration the source of electrons is from _____.

11. Suppose that an animal inhales a radioactive isotope of oxygen whose location can be traced in the animal. Where would the radioactive oxygen atoms eventually show up?

12. Fill in the following comparison chart of cellular respiration and photosynthesis by using the words: "consumed" and "produced."

	Photosynthesis	*Cellular respiration*
Oxygen		
Carbon dioxide		
Glucose		

Chapter 5
DNA, GENE EXPRESSION, AND BIOTECHNOLOGY—
WHAT IS THE CODE AND HOW IS IT HARNESSED?

Learning Objectives

- Understand the structure and function of DNA
- Explain the relationship between genes and proteins
- Describe the processes of transcription and translation
- Identify the impact and causes of mutations
- Explain how genes with mutations can cause illness or disorder
- Understand the basic tools used in the application of biotechnology
- Explain how biotechnology can help produce pharmaceuticals, treat disease, and even prevent disease
- Define terms such as genetic engineering, recombinant DNA technology, PCR, and transgenic organisms
- Identify the strengths and weaknesses of gene therapy

Chapter Outline

I. DNA: What Is It, and What Does It Do?

A. Social Impact of DNA

- Briefly explain how the biological molecule, DNA, has the potential to impact the criminal justice system in our society.

B. The DNA Molecule Contains the Instructions for the Development and Function of All Living Organisms

- Two important discoveries about DNA include the understanding that:
    - DNA is passed down from generation to generation, and
    - DNA is an encoded set of instructions for controlling growth and development

- Besides James Watson and Francis Crick, list two important scientists involved in DNA research and describe their accomplishments.

    1.

    2.

- DNA stands for _____ and is one of the important _____, or macromolecules that contain genetic information.

- DNA is composed of individual building blocks called _____.
    o List the components of this building block.

        1.

        2.

        3.

- On the figure to the right, complete the following:

    o Highlight or circle a single nucleotide.
    o In a different colored writing utensil, circle and label the parts of the "uprights" or "backbones" of the double helix.
    o Circle and label the part of the "rungs" of the double helix.

- Make a list of the four different nitrogen-containing bases:

    o Which bases will always be a base pair, or bond together?

    o What holds the base pairs together?

C. Genes Contain Instructions for Making Proteins

- Just like the sequence of letters in a word is crucial to the meaning of the word, the sequence of the DNA bases is critical to the "language" of DNA. Explain why the bases of DNA make up a "secret code" (of sorts) and what information this "code" contains.

- Define **genome**:

- Complete the chart below in regard to the organism's genome.

	Eukaryotic organisms	Prokaryotic organisms
Where is it located?		
How is it organized?		

- Humans have _____ (number) **chromosomes**, which can be arranged into _____ (number) pairs.

- Can we assume that an organism with fewer chromosomes than a human is less complex? Explain using an example.

- Explain the relationship between a **gene** and a chromosome.

  o An allele is:

  o Different alleles can specify different **traits**. Define the term **trait** and give an example.

D. Not All DNA Contains Instructions for Making Proteins

- The size of the genome varies from organism to organism and different species have different amounts of DNA. Just as the number of different chromosomes does not parallel the "complexity" of an organism, neither does the amount of DNA.

- What percentage of DNA in humans consists of genes?

  o Some biologists refer to the remainder of the DNA as _____.

- Non-coding DNA, or DNA not used in protein production, that is found within genes is called _____.

- Which type of organism has more non-coding DNA: bacteria or eukaryotes? Explain.

- Do scientists know the purpose of non-coding DNA? Explain.

E. How Do Genes Work?

- Think about the number of light bulbs in your house. Are they all turned on at the same time? Hopefully not! You may use some more than others, and some you may never use. Just like your house, your cells have all of the genes necessary to produce all of the protein in the body; however, not all protein is produced by every cell.

- Can you tell what someone's genotype is just by looking at them?   In your explanation include a description of an organism's genotype.

- Can you observe an individual's phenotype when you walk into a room? Explain. Include a definition of phenotype and give an example.

- List, and give a brief description of the two main steps that allow the cellular instructions, or the genes, to provide the information to produce a protein.

    1.

    2.

        o   What molecule is necessary to allow the instructions, or the **code**, to be relayed properly when producing a protein?

II. Building Organisms: Information in DNA Directs the Production of the Molecules that Make up an Organism

- An overview of the process of how proteins are produced from genes:
  - In the first step, transcription, a copy or transcript called
    _____ is created from different ribonulceotides
    by the enzyme _____.
  - The protein factories, the _____, must be available. The protein
    is built with amino acids that are carried to the ribosome by the transport
    molecule _____.

A. Transcription

- In about one sentence, use your own words to describe the main goal of transcription.

- Including the important molecules that are involved, describe the four steps of
  transcription below.

  1. Recognize and bind

  2. Transcribe

     - If the DNA base sequence was AGCTACATG, indicate the mRNA transcript
       produced.

  3. Terminate

  4. Capping and editing

- Summarize transcription by completing the chart below.

Location in the cell	
Product produced	
What must happen after the product is produced before moving on to the next step?	
Where, in the cell, does the product go next?	

B. Translation

- In one sentence, use your own words to describe the main goal of translation.

- Besides the mRNA transcript, what are the other three molecules necessary in order for translation to occur? List them and provide a brief description of their job.

    1.

    2.

    3.

- One important aspect of translation is ensuring that the amino acids end up in the correct sequence when building the protein. The tRNA plays an important role in this step. The tRNA transports a specific _____, which is attached to one end of the molecule. The other end of the tRNA molecule is used as an attachment site. The attachment site is a set of three bases that will pair with a set of three bases in the mRNA strand. These three bases in the mRNA strand are also called a

    _____.

- Label the parts of figure below.

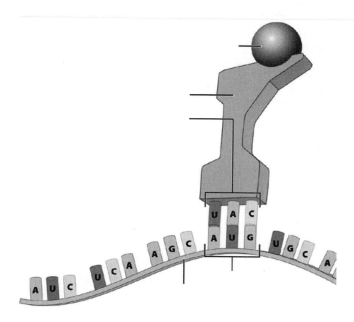

- Briefly explain each step of translation:

    1. Recognize and initiate protein building

    2. Elongate

    3. Terminate

- Describe what happens to the mRNA strand after translation is complete.

C. Genes Are Regulated in Several Ways

- In addition to the processes of transcription and translation, an important aspect of protein production is how genes are regulated. Briefly define:
  - o **Gene expression:**

  - o **Gene regulation:**

- What is a microarray and how would it be used to study gene expression?

- Explain the three different elements of an **operon** and how they work.
  1.

  2.

  3.

- Give an example of how transcription rates can be increased and how transcription can be blocked or decreased in eukaryotic genes.

III. Damage to the Genetic Code Has a Variety of Causes and Effects

A. What Causes a Mutation, and What Are Its Effects?

- If there is a change to the sequence of DNA within a coding region that alters the DNA bases, it is called a _____.
  - o If the sequence of DNA is the instructions to produce a specific protein, then a change to those instructions will result in a change to the protein produced.

- Since the word mutation has a negative connotation, we expect the changed protein to have a negative effect. While the protein will be changed and could also be nonfunctional or detrimental, the results will vary. A mutation can have:
    1. a detrimental effect on the organism,
    2. no effect on the organism, or even
    3. a positive impact for the organism.

    o Give an example of a mutation within a gene and its impact.

- Can mutations occur in gametes or non-sex cells or both? Explain.

- Two major types of mutations include **point mutations** and **chromosomal aberrations**. Briefly describe each type of mutation and include the likely impact of the mutation.

    1. Point mutations

    2. Chromosomal aberrations

- What are the potential causes of mutations in our cells' DNA? List the three major causes of mutations and give an example.

    1.

    2.

    3.

B. Faulty Genes, Coding for Faulty Enzymes, Can Lead to Sickness

- Genetic disease is often the result of a mutation in a DNA sequence that codes for a protein that is commonly an _____. If the non-functional protein cannot do its job, then it will impact the functioning of the cells and possibly result in illness.

5. Identifying bacterial colonies that have received the gene of interest

- o   Describe **hybridization**.

- o   What is a **DNA probe**?

B. Biotechnology Can Improve Food Nutrition and Make Farming More Efficient and Eco-friendly

- Selecting desirable traits and breeding to produce better food has been a practice in agriculture for a very long time. It was an imprecise form of **genetic engineering**. Today, with **recombinant DNA technology**, there are more sophisticated ways to utilize biotechnology in agriculture.

- Instead of simply hoping the offspring will express the desired trait selected for after generations of breeding, the use of recombinant DNA technology can allow for a trait from a completely different species to be expressed.

- Crops or animals that are produced using recombinant DNA technology are referred to as _____ or abbreviated _____.

- One example of a crop produced from recombinant DNA technology is a type of rice referred to as _____.

  - o   What are the consequences of vitamin A deficiency?

  - o   What do mammals make vitamin A from?

  - o   In what area is this vitamin deficiency most problematic?

  - o   Briefly explain the process for producing golden rice, including the source(s) of the gene(s).

- What percentage of certain crops, such as cotton and soybeans, are genetically modified?

  - o   In general, why is this number so high?

- Briefly explain the two specific reasons there has been such an expansive adoption of genetically modified crops.

  1.

  2.

- Utilizing recombinant DNA technology has been most successful in providing insect and herbicide resistance for crops, as well as more rapid growth to market size of fish, specifically salmon.

  o Explain how bacteria have been used to engineer corn to be resistant to insects.

  ▪ How do crops resist the effects of herbicides?

  o Explain the benefits of transgenic salmon.

  ▪ What are some of the concerns surrounding transgenic salmon?

C. Fears and Risks: Are Genetically Modified Foods Safe?

- List and explain at least three concerns that surround the production and use of genetically modified foods.

- Many of those opposed to GM foods try to make a point that they are not natural. Why is this a weak argument?

- Give an example of why it is important to balance the risks with the benefits when producing or consuming genetically modified foods.

V. Biotechnology Has the Potential for Improving Human Health (and Criminal Justice)

A. Treatment of Diseases and Production of Medicine

- What is diabetes?

- How was it treated prior to the early 1980s?

- How has recombinant DNA technology been used to help treat diabetes more efficiently?

- **Human growth hormone** (HGH) and **erythropoietin** are two important examples of recombinant DNA technology applications.

    o   Describe the effect of each hormone on the body.

    o   How has each of these hormones been involved in controversy?

B. Gene Therapy

- The text describes three scenarios or points in time when biotechnology has the potential to help prevent disease. Describe each of these three situations, or the question that is seeking an answer, and explain how biotechnology could play a role.

- Describe the goal of **gene therapies**.

- **SCID** stands for _____.
  How does gene therapy have the potential to cure this genetic disorder?

- Explain four potential problems with gene transfer for gene therapy.

C. Cloning

- While the term cloning evokes images of sci-fi movies, there are actually different meanings of the word cloning. With DNA technology it is possible to:
    - clone an entire organism (Dolly)
    - clone tissues
    - clones genes

- Using the graphic below, outline the steps involved in producing an organism identical to that of a donor cell.

Testing and Applying Your Understanding

Multiple Choice (For more multiple choice questions, visit www.prep-u.com.)

1. Which of the following nucleotide bases are present in equal amounts in DNA?
    a) adenine and cytosine
    b) adenine and guanine
    c) adenine and thymine
    d) thymine and guanine
    e) thymine and cytosine

2. The central dogma of molecular biology states that:
    a) RNA is transcribed into protein.
    b) DNA is transcribed into RNA.
    c) DNA is translated into protein.
    d) DNA is transcribed into RNA, which is translated into protein.
    e) DNA is translated into RNA, which is transcribed into protein.

3. PCR is a common technique used to amplify large portions of specified gene sequences. What does the acronym PCR stand for?
    a) perform correct response
    b) purification cleansing reagent
    c) Paul Carl Rease
    d) polymerase chain reaction
    e) powerful catabolize reaction

4. The technique often used in forensics that identifies individuals based on their genetic differences is referred to as:
    a) DNA cloning.
    b) DNA screening.
    c) DNA typing.
    d) DNA fingerprinting.
    e) DNA analyzing.

5. When a triplet of bases in the coding sequence of DNA is GCA, the corresponding codon for the mRNA that is transcribed from it is:
    a) CGT.
    b) GCU.
    c) GCT.
    d) UGC.
    e) CGU.

6.   What are the three functions of the tRNA molecule?
a) The tRNA molecule carries an amino acid, associates with rRNA molecules, and binds to one of three sites on the large subunit of an mRNA molecule.
b) The tRNA molecule transcribes DNA, associates with rRNA molecules, and synthesizes activating enzymes.
c) The tRNA molecule carries an amino acid, associates with mRNA molecules, and binds to one of three sites on the large subunit of a ribosome.
d) The tRNA molecule carries an amino acid, associates with mRNA molecules, and replicates DNA.
e) The tRNA molecule transcibes, translates, and replicates the DNA.

7.   The expression of a gene to form a polypeptide occurs in two major steps. What are these two steps in their correct chronological order?
a) transcription and then translation
b) replication and then transcription
c) transcription and then replication
d) translation and then transcription
e) replication and then translation

8.   Gene therapy involves:
a) introducing non-defective genes into the cells of an individual with a genetic disorder.
b) drug treatment of patients with genetic disorders at specific times that correspond with cell division.
c) the replacement of organs from patients with genetic disorders by transplant.
d) no controversial or ethical questions.
e) All of the above are correct.

9.   Which of the following would be considered a transgenic organism?
a) a rat with rabbit hemoglobin genes
b) a bacterium that has been treated with a compound that affects the expression of many of its genes
c) a fern grown in cell culture from a single fern root cell
d) a human treated with insulin produced by *E. coli* bacteria
e) All of the above are correct.

10.   The genes that code for proteins and the genes for RNA products such as rRNA and tRNA constitute a surprisingly small portion of the genomes of most multicellular eukaryotes. The majority of most eukaryotic genomes consist of non-coding regions, sometimes described as "junk DNA." However, recent evidence shows that even this so-called "junk DNA" can play important roles. Which of the following is NOT a type of non-coding DNA?
a) introns
b) transposable elements
c) regulatory sequences
d) repetitive DNA
e) exons

Short Answer

1.   In thinking about the properties of chemical bonds, explain why it is beneficial that DNA base pairs are held together with hydrogen bonds.

2.   Suppose your cells were deficient in amino acids. Would this deficiency have a greater impact on transcription or translation? Explain why.

3.   Explain what happens to the intron regions of DNA after transcription. Why is this important to the cell and the process of protein production?

4.   Compare the impact of a mutation in the DNA sequence of a gene with the impact of a mutation in the intron region of DNA.

5.   Describe the differences between a point mutation and chromosomal aberrations.

6. Biotechnology includes tools that involve DNA technology. These tools can be used to detect disease-causing genes in an effort to prevent or alter the course of disease. Explain an ethical dilemma that could arise as a result of utilizing this technology.

7. Describe, using an example, a genetically modified (GM) food.

8. List one fear or concern you had in regard to GM foods. Was that concern addressed in this section? If not, do a little investigating on the topic and explain here.

9. Cloning transgenic animals—such as sheep, pigs, or cattle—can be beneficial to human health. What is a transgenic animal and why would having several clones be valuable?

Chapter 6
CHROMOSOMES AND CELL DIVISION—CONTINUITY AND VARIETY

Learning Objectives

- Identify different types of cell division and their purpose(s)
- Explain the process by which prokaryotic cells divide
- Describe the major phases of the cell cycle
- Describe the process of DNA replication
- Define the purpose and steps of mitosis
- Explain how the cell cycle and cancer are related
- Define cancer
- Explain how cancer can be treated
- Explain the steps of meiosis
- Compare and contrast meiosis in males versus females
- Compare and contrast mitosis and meiosis
- Understand the differences in asexual and sexual reproduction
- Explain the impact of having an abnormal number of chromosomes

Chapter Outline

I. There are Different Types of Cell Division

A. The Basics of Cell Division and Chromosomes

- It is imperative that our cells, such as skin cells, divide in order to replace cells that have died. However, it is also important the cells divide only a limited number of times as immortal cells present problems such as cancer.

- A protective cap located at the end of the DNA, called a _____, acts to limit the number of cell divisions an individual cell undergoes. Explain how this "cap" can protectively limit cell divisions.

  o This protective cap can allow for about _____ (number) of cell divisions from birth.

- Different cell types undergo different processes to complete cell division. Summarize the types of cell division by completing the chart below. Include all methods.

Cell type	Method(s) of division
Prokaryotes	
Eukaryotes	

- Define a chromosome.

    1. Describe the chromosome(s) of prokaryotes; include shape, size, and number.

    2. Describe the chromosome(s) of eukaryotes; include shape, size, and number.

        - In addition to DNA, the eukaryotic chromosomes have a specialized protein called a _____. DNA is wrapped around this protein in order to package the chromosomes in the nucleus.

B. Binary Fission

- A single prokaryotic cell is able to divide into two via the process of **binary fission**.

- This type of reproduction is termed _____ reproduction and is different from **sexual reproduction** because:

- Binary fission begins with a single _____ cell (the original cell). The cell must undergo replication. Describe what occurs in replication.

- After replication is completed, the cell elongates and pinches into two _____ cells. Describe what each resulting cell contains.

C. Cell Cycle

- To understand the process of eukaryotic cell division, it must be put in context of the **cell cycle**. Describe the purpose of this cycle.

- Different cells in our body will undergo a different process of division depending on their cell type. Two major categories of cell types in the body are somatic and reproductive cells.

    o Describe a **somatic cell**.

    o Describe a **reproductive cell**.

        ▪ These cells are also referred to as _____.

- Phases of the Cell Cycle

    Explain what is occurring in the cell during each step of the cell cycle.

    1. Interphase

        a. Gap 1

        b. DNA synthesis

        c. Gap 2

    2. Mitotic phase

        a. Mitosis

        b. Cytokinesis

- There are checkpoints between phases of the cell cycle to ensure the cellular environment is suitable for division.

D. Replication

- From the steps in the cell cycle, you now know that the cell must duplicate all of the cellular contents prior to dividing. Arguably, the most important item to copy is the cellular genetic material.

- Key to replication of DNA is its feature of **complementarity**. Explain what it means to say DNA strands are complementary.

- The basic process of DNA replication consists of the DNA molecule unwinding and unzipping and a new strand of DNA is then synthesized. Specifically this process occurs in two major steps. List the steps below and explain what occurs during that step.

    1.

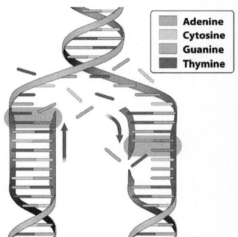

Adenine
Cytosine
Guanine
Thymine

    2.

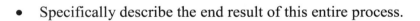

- Specifically describe the end result of this entire process.

- Mistakes are possible during the process of replication which would result in a type of mutation. What are the possible outcomes of that "mistake"?

II.   Mitosis Replaces Worn-Out Old Cells with Fresh New Duplicates

A. Purpose and Overview

- In your own words, give the purpose of mitosis.

- Mitosis is required for:

    o **Growth**
        - When organisms experience growth their cells don't get bigger; cell division occurs to create new cells.

    o **Replacement**
        - Give an example of a cell type that is frequently replaced and how often replacement occurs.

        - Define **apoptosi**s and explain why it is a positive thing for the organism.

- What types of cells perform mitosis? What are the exceptions?

- For cell division to occur the parent cell must create a _____ of each chromosome in interphase. Then, all the other cellular components are replicated and the parent cell divides into two _____ cells.

- The process to divvy up the cellular material occurs in _____ (number of) specific, regulated steps.

B. Steps of Mitosis

- What is the end goal of this process?

- **Interphase**

    o During the _____ phase of interphase, a copy of each chromosome is made.

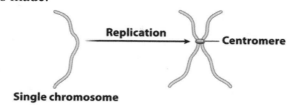

        ▪ Define a **chromatid**

        ▪ On the graphic above, label the **sister chromatids**

    o This process will allow for each daughter cell to have an identical set of genetic material.

- **Mitosis**

    Below, list each step of mitosis and major events that occur in that step.

    1. Prophase:

        o What are **spindle fibers** and what role do they play?

    2.

    3.

    4.

- After telophase, the final task that must occur is the division of the cell's cytoplasm, also referred to as _____, to result in two separate cells.

- Are the two resulting daughter cells identical? Are they identical to the parent cell? Explain.

- Create a mnemonic device to help you remember the order of the steps of mitosis.

C. Cancer

- The process of cell division, in part, is regulated by the individual phases of the cell cycle. The division of the cell's DNA is further regulated by the four individual steps that make up the phases of mitosis. These carefully regulated steps help to ensure that the process of cell division is occurring properly and only when necessary. Unrelated cell division can lead to _____.

    o Define **cancer**.

- A change or alteration of DNA can lead to uncontrolled cell growth. List some possible causes for this sort of change or alteration.

- How are cancer cells different from normal cells? List and explain three important differences.

    1.

    2.

    3.

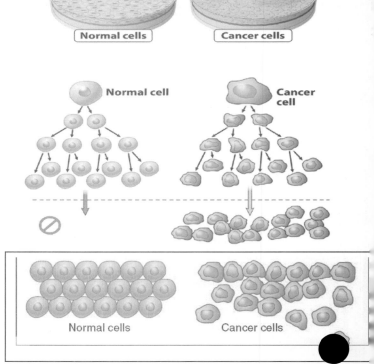

- Define a **benign tumor**; be specific.

- Define a **malignant tumor**; be specific.

  o   What is metastasis?

- Cancerous cells are often described as invasive. How does their invasiveness affect surrounding tissues and/or organs?
- There are two modalities or mainstream methods of treating cancer. They are:

  1.

  2.

- Define **chemotherapy**.

- Define **radiation**.

- Why are there difficult and harmful side-effects as a result of these treatments?

  o   List specific cell types that are affected by either chemotherapy, radiation, or both.

  o   What do these cell types have in common?

- What is the chemical resveratrol an example of?

III. Meiosis Generates Sperm and Eggs and a Great Deal of Variation

A. Overview

- The second method of cell division for eukaryotes is **meiosis**. Meiosis is cell division for the purposes of sexual reproduction. Sexual reproduction results in offspring by the union of two unique reproductive cells, or **gametes**, in the process _____.

- What would happen if mitosis was the process of cell division that produced gametes? Explain.

- One important aspect of meiosis is that the process results in cutting the organism's chromosome number in half. In humans this means a gamete has _____ (number of) chromosomes.

    o Haploid refers to a cell that:

    o Diploid refers to a cell that:

- In fertilization, two _____ (haploid or diploid?) cells unite to produce a _____ (haploid or diploid?) cell.

- The two goals of meiosis are:

    1.

    2.

B. Steps of Meiosis

- The purpose of meiosis is to produce _____. It occurs in specialized _____ (diploid or haploid?) cells found in the **gonads,** or the _____ in females and the _____ in males.

- The cells' 46 chromosomes (for humans) can be grouped into _____ (number of) pairs. Each pair contains one paternal copy and one maternal copy. In order to identify this type of pair, they are referred to as _____ pairs.

- Just as with mitosis, DNA replication must take place in interphase before the process of cell division begins. Therefore, each homologue has an identical copy, or there are 46 pairs of sister chromatids before meiosis begins.

- The parent cell must undergo two rounds of division to complete meiosis.

    o In the first round, or meiosis I, what happens?

    o In the second round, or meiosis II, what happens?

    o The end result is _____ (number of) daughter cells that are each _____ (haploid or diploid?).

- **Meiosis I**

    List the major events in each phase of Meiosis I.

    1. Prophase I:

        o What is **crossing over**?

    2. Metaphase I:

    3. Anaphase I:

    4. Telophase I and cytokinesis:

- **Meiosis II**

    List the major events in each phase of Meiosis II and highlight how the phase is different than in Meiosis I.

    1. Prophase II:

    2. Metaphase II:

    3. Anaphase II:

4. Telophase II and cytokinesis:

- Characterize the four resulting daughter cells, including the chromosome number and if they are identical to each other or the parent cell.

C. Gamete Production in Males versus Females

- The end result of meiosis in both males and females are reproductive cells, or gametes. However, in males these cells eventually become _____ , and in females these are _____.
    o Describe, specifically, how the gamete differs between males and females.

- While both are used in reproduction, the different gametes are a reflection of meiosis occurring differently in males and females.

    o Describe the four cells that are the result of meiosis in males.

    o Describe the four cells that are the result of meiosis in females.

- Why is it advantageous that the egg carry extra cytoplasm?

D. Diversity as a Result of Crossing Over

- When does crossing over occur?

- Crossing over results in a different combination of alleles. Explain how.

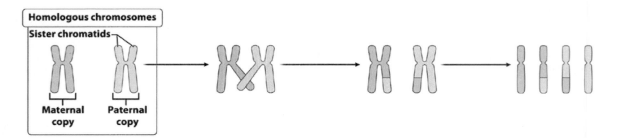

- How does crossing over result in variation or diversity with in a species?

E. Sexual versus Asexual Reproduction

- Complete the chart to highlight the advantages and disadvantages of reproduction via mitosis and asexual reproduction as well as meiosis and sexual reproduction.

	Advantages	Disadvantages
Sexual reproduction	(be sure to include the sources of genetic variation here)	
Asexual reproduction		

IV. There Are Sex Differences in the Chromosomes

A. Sex Determination in Humans

- There are many old wives' tales outlining ways to predetermine the sex of a baby. However, we know that the child's sex is determined by the combination of sex chromosomes—specifically, the one inherited by the child's _____.

- Human sex chromosomes are the _____ and the _____ chromosomes.

- The combination of sex chromosomes a male carries is _____ and the combination a females carries is _____.

- How are these chromosomes different than the other 22 pairs of chromosomes? Include what type of information or instructions they carry.

- Describe the resulting sex chromosome found in each egg. Describe the resulting sex chromosome found in sperm cells. Are they the same or different? Explain.

- What are some of the physical differences between the X and the Y chromosomes?

B. Sex Determination in Other Species

- Define **hermaphrodite** and give an example of an organism that could be classified as a hermaphrodite.

- Explain how the following species, each with male and female individuals, determine the sex of their offspring.

    o   Birds:

    o   Bees:

    o   Turtles:

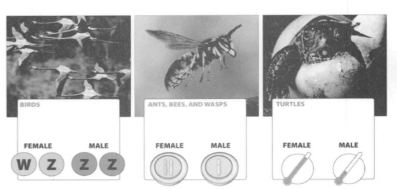

V. Deviations from the Normal Chromosome Number Lead to Problems

A. Karyotype: Determining Chromosome Number

- What is one reason the chances of having a child with a chromosomal abnormality, or a disorder related to an incorrect number of chromosomes, increases as the mother's age increases?

- What is a **karyotype**, and what is it used for?

- Describe the five steps involved in producing a karyotype.

    1.

    2.

    3.

    4.

    5.

- If you wanted to have a karyotype created of your genetic make-up, where would the cells likely come from?

- Two methods for collecting cells from a developing fetus include an amniocentesis and chorionic villus sampling. Briefly describe what is involved in each procedure.

    1. Amniocentesis:

    2. Chorionic villus sampling (CSV):

- Gametes with an abnormal number of chromosomes likely were the result of a "mistake" in the process of meiosis. This unequal distribution of genetic material is called _____.

    o   When can this occur in meiosis?

    o   A gamete with an extra chromosome is called a _____.

    o   A possible resulting condition is an individual with an extra copy of chromosome 21. This condition is commonly called

    _____.

- Why are more chromosomal abnormalities associated with the mother versus the father?

B. Abnormal Number of Sex Chromosomes

- Missing a non-sex chromosome or having an extra non-sex chromosome is most likely fatal. However, this is not the case with an abnormal number of sex chromosomes.

- Provide the genotype and the syndrome characteristics for the following syndromes.

    1. Turner syndrome:

    2. Klinefelter syndrome:

    3. XYY males:

    4. XXX Females:

- Why is an individual with Klinefelter syndrome *not* a hermaphrodite?

Testing and Applying Your Understanding

Multiple Choice (For more multiple choice questions, visit www.prep-u.com.)

1. During which phase of the cell cycle is DNA replicated?
    a) interphase
    b) prophase
    c) metaphase
    d) anaphase
    e) telophase

2. Which step in meiosis is responsible for generating genetic diversity?
    a) metaphase of Meiosis I
    b) anaphase of Meiosis I
    c) telophase of Meiosis I
    d) prophase of Meiosis I
    e) None of the above; genetic diversity is generated during Meiosis II.

3. Prokaryotic cells can divide via:
    a) meiosis.
    b) mitosis.
    c) binary fission.
    d) All of the above.
    e) None of the above.

4. Mitosis results in:
    a) diploid gametes.
    b) somatic cells with twice as much genetic material and a unique collection of alleles.
    c) haploid gametes.
    d) somatic daughter cells with the same number and composition of chromosomes.
    e) four daughter cells.

5. A karyotype reveals:
    a) the autosomes but not the sex chromosomes.
    b) 23 pairs of chromosomes.
    c) the shape of the spindle apparatus.
    d) the number, shapes, and sizes of chromosomes in an individual cell.
    e) the sex chromosomes but not the autosomes.

6. Nondisjunction:
    a) is the cause of sex determination in birds and mammals.
    b) occurs during mitosis but not meiosis.
    c) is the division of cytoplasmic constituents.
    d) is the unequal division of the genetic material during cell division.
    e) is the exchange of genetic material between the chromatids from homologous chromosomes.

7. Which of the following does NOT occur during Prophase I of meiosis?
   a) Chromosomes begin to condense.
   b) Spindle microtubules form.
   c) Homologous pairs of chromosomes are aligned at the metaphase plate.
   d) A protein structure called a synaptonemal complex forms between the homologues.
   e) Crossing over between tetrads occurs.

8. In humans, the haploid number n equals:
   a) 46.
   b) $2n$.
   c) 44.
   d) n.
   e) 23.

9. Somatic cells can include:
   a) kidney cells.
   b) heart cells.
   c) sperm cells.
   d) All of the above are correct.
   e) Both a) and b) are correct.

10. A 30-year-old woman has a 1 in 3000 chance of giving birth to a child with trisomy 21, but a 48-year-old woman has a 1 in 9 chance. Which of the following is the most likely explanation for why a woman has a higher probability of giving birth to a child with trisomy 21 (Down syndrome) as she ages?
    a) Older women have older oocytes, and the older the oocytes, the greater the chance they will experience a nondisjunction event during meiosis.
    b) Older women have lived longer and have an increased percentage of toxins and oxidative damage in their body. Since these are the major causes of trisomy 21, its likelihood increases.
    c) Older women are more likely to mate with older men, and the older the man, the older the sperm. Older sperm are more likely to experience a nondisjunction event during meiosis.
    d) Older women have older oocytes, and the older the oocytes, the greater the chance they will experience failure of cytokinesis during cell division.
    e) Answers a) and d) provide the most likely explanations.

Short Answer

1. Normal cells are limited in the number of times they are able to divide and continue operating as a functional cell. Explain how cells "know" how many times they have divided.

2. Explain why cell division by mitosis would not work for the purpose of producing gametes.

3. Boxers are a breed of dogs prone to developing tumors. Concerned about a lump on her dog's hindquarters, Amy brings her dog to the vet. What cellular characteristics would the vet look for to determine if the tumor was benign or malignant?

4. The end result of meiosis differs between males and females. Explain the major differences between the two and what might be advantageous about these differences.

5. A young couple trying to have a child seeks medical intervention after years of being unsuccessful. They discover the man has Klinefelter syndrome.
   a. How did this occur? Explain in the context of the process of meiosis.
   b. He is upset and wants to blame his situation on one of his parents. While obviously his parents did not intend for him to have this condition, who, biologically, might he be able to "blame" and why?

6. You are examining a picture of a human cell in the process of dividing. You can see 46 "X"s that are lined up in single file down the center of the cell.
    a.   What stage of cell division is represented in the picture?
    b.   Explain why the chromosomes are described as an "X."

7. Explain how the human sex chromosomes differ from each other and why it is more significant to have an extra non-sex chromosome.

8. Could a karyotype detect mutations in a specific sequence of DNA? Explain why or why not.

9. Compare the process of cell division and reproduction in bacteria versus eukaryotic cells.

Chapter 7
MENDELIAN INHERITANCE

Learning Objectives

- Explain the concept of a single-gene trait
- Describe Mendel's contributions to the field of genetics
- Be able to define the terms: gene, allele, dominant, recessive, homozygous, and heterozygous
- Characterize the difference between an organism's genotype and its phenotype
- Demonstrate the ability to perform Punnett squares to predict the offspring of parents with particular genotypes
- Understand how to use the rules of probability to predict the inheritance of specific traits
- Explain how a test-cross can be used to determine the genotype of an organism.
- Predict patterns of inheritance based on pedigree analysis
- Explain how each of the following genetic "rules" works: incomplete dominance, codominance, multiple alleles, polygenic inheritance, and pleitropy
- Understand how the ABO and Rh markers contribute to human blood type
- Explain how blood type compatibility is determined
- Describe how sex linked traits are inherited
- Understand that phenotypes are a combination of genotypes and the environment
- Explain why linked genes do not assort independently

Chapter Outline

I. Why Do Offspring Resemble Their Parents?

- In Chapter 5 you learned about DNA and in Chapter 6 you learned about meiosis and the production of gametes. A few points to recall that are pertinent to our discussion of genetics are the following:

    o In a diploid cell, there are _____ copies of each chromosome present.

    o Each human diploid cell has a total of _____ chromosomes or _____ pairs of chromosomes

    o A copy of a gene that can be passed from a parent is termed an

    _____.

A. Single-Gene Traits

- The passing of characteristics from parents to offspring via genes is referred to as

    _____.

- The inheritance of some traits is easy to predict. When a single gene influences a trait (a **single-gene trait**), will the inheritance of that trait be relatively easy or difficult to predict? Justify your answer.

- Provide an example of a human trait that would **not** be considered a single-gene trait:

B. Mendel's Research Design

- Gregor Mendel's work was fundamental to all modern genetic theories. There were three features of Mendel's work that were particularly important to his success. Those were:

    1.

    2.

    3.

C. Mendel's Law of Segregation

- Mendel's experiments in crossing pea plants led to predictable results. For example, crossing **true-breeding** purple pea plants to white pea plants always led to purple offspring. This is because purple color is **dominant** while white color is **recessive**.

- The concept of dominance is often misunderstood. People tend to think that dominant alleles are more common than recessive alleles or that they are in some way more advantageous than recessive alleles.

  o In your own words, describe how a dominant allele is different from a recessive allele.

- For any given trait, an individual receives exactly _____ copy of a gene from each parent via the gamete. This leads to individuals whose diploid cells have exactly _____ copies of each gene.

- When an individual inherits two of the same allele from both parents, we call this _____ and when an individual inherits two different alleles from their parents, we call this _____.

- Briefly summarize Mendel's **law of segregation**.

D. Genotypes, Phenotypes, and Punnett Squares

- Explain what the term genotype means.

- Explain what the term phenotype means.

- What is a Punnett square used for and how would you go about setting one up?

- In the following example, there are two possible alleles for the gene that determines whether a person has dimples. Having dimples is dominant and the allele will be represented as D. Not having dimples is recessive and the allele will be represented as d.

- For each possible genotype, fill in the corresponding phenotype.

Possible allelic combinations	Genotype	Phenotype
DD	Homozygous dominant	
Dd	Heterozygous	
dd	Homozygous recessive	

- Suppose your roommate has dimples. Can you be sure of her genotype? Explain your answer.

- Show a Punnett square for the dimples trait between a homozygous recessive parent and a heterozygous parent. Indicate the phenotypic ratio for the offspring.

- In the chapter you saw examples of Punnett squares using albinism in giraffes. Given the following example, fill in the gametes that would be produced by each parent, the offspring that would be produced, and the phenotypic ratio of the offspring.

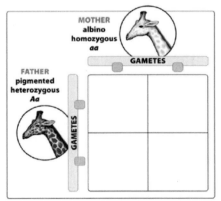

II. Probability and Chance Play Central Roles in Genetics

- There are two reasons that the rules of probability are used in genetics. Those reasons include:

1.

2.

- What would be the probability of a child inheriting Tay-Sachs disease from a mother who is heterozygous for the trait and a father who is homozygous dominant for the trait? Explain how you would use the rules of probability to determine your answer.

A. Test Crossing

- During a **test-cross**, an organism with a _____ phenotype and an unknown genotype is always bred with a mate that is _____.

    o If half of the offspring of the test-cross show the dominant phenotype and half show the recessive phenotype, then the genotype of the unknown parent was_____.

    o If all of the offspring of the test-cross show the dominant phenotype, then the genotype of the unknown parent was_____.

B. Pedigree Analysis

- **Pedigrees** are useful tools for determining the pattern of inheritance for a specific allele. Within a pedigree, females are represented by _____ and males are represented by _____. How can you tell which individuals in a pedigree contain a trait of interest?

- If a trait of interest is located on an **autosome**, the pattern of inheritance will be _____ or _____.

- If a trait of interest is located on the X or Y chromosome, the pattern of inheritance will be _____.

- If a pedigree shows a **carrier** for a specific trait, what does this mean?

III. The Translation of Genotypes into Phenotypes Is not a Black Box

- Not all genetic traits operate by simple dominance of one allele over a recessive allele. There are other genetic rules that govern the inheritance of specific genetic traits.

A. Incomplete Dominance and Codominance

- In some cases one allele does not clearly dominate another. In this case, individuals with a _____ genotype express a unique phenotype.

- At first glance, **incomplete dominance** and **codominance** seem very similar to each other. Explain how they are different.

- Roan cows have an interesting coat color. The have blotches of red fur and blotches of white fur. If you were told that their genotype is heterozygous, which rule of inheritance do you think applies to the development of roan coat color? Explain your answer.

B. Multiple Alleles and Blood Type

- When a single gene has more than two alleles, this is referred to as **multiple allelism**. Human blood type is an example. How many alleles are involved in this trait? How many alleles do you inherit for this trait?

- How does the term antigen relate to blood type?

- In the chart below, indicate the four human blood types, the potential genotypes of each, and the antigens that would be present on the red blood cells of an individual with that specific blood type:

Possible Blood Types	Potential Genotypes	Antigens Coded For

- During blood transfusions, compatibilities between donor and recipient must be accounted for. Complete the following table from Figure 7-21:

BLOOD TYPE	CAN DONATE TO	CAN RECEIVE FROM
Type A • Has A antigens • Produces antibodies that attack B antigens		
Type B • Has B antigens • Produces antibodies that attack A antigens		
Type AB • Has A and B antigens • Produces neither A nor B antibodies • Universal recipient		
Type O • Has neither A nor B antigens • Produces antibodies that attack A and B antigens • Universal donor		

- Speaking in terms of the immune system, what would happen if a person with type A blood were to receive a type B blood transfusion?

- In addition to the ABO markers, there is another genetic marker that determines whether an individual's blood type is positive or negative. This is referred to as the

  _____.

- What happens when a person with Rh+ blood donates to someone with Rh- blood?

C. Multigene Traits

- **Polygenic traits** are ones in which a single phenotype is influenced by _____ genes.

- When alleles from multiple genes influence a single phenotype, this is referred to as
  _____.

- Provide three examples of human traits that are polygenic:

- In **pleitropy**, a single gene influences multiple traits. Some examples of this would
  include:

D. Sex-Linked Traits

- Traits that are coded for on sex chromosomes have unique patterns of inheritance. Recall
  that the genotype of a female is _____ and the genotype of a male is _____.

- Which gender is more likely to exhibit a sex-linked recessive trait?

- Suppose that a female exhibits red-green color-blindness. What must the genotypes of her
  parents have been? Explain your answer and draw a Punnett square.

E. Environmental Effects

- It is important to note that traits are not shaped by genotype alone. The ultimate
  phenotype expressed is shaped by both the genes inherited in combination with the
  _____.

- Provide three examples of human phenotypes that can be shaped by the factors other than
  genes:

    1.

2.

3.

IV. Some Genes are Linked Together

- Mendel's **law of independent assortment** can be used to explain how traits are inherited independently of each other.

- In order for independent assortment to occur, the traits must exist on separate

  _____.

- When traits are considered **linked**, they exist in close proximity on the same chromosome. How can meiosis be used to explain why linked traits do not assort independently of one another?

Testing and Applying Your Understanding

Multiple Choice

1. Which of the following statements about dominant traits is correct?
    a) They are observed less frequently than recessive traits.
    b) They are observed more frequently than recessive traits.
    c) They are observed one-quarter as frequently as heterozygous traits.
    d) They increase in frequency over evolutionary time.
    e) None of the above is correct.

2. In certain plants, red flowers are dominant to white flowers. If a heterozygous plant is crossed with a homozygous red-flowered plant, what is the probability that the offspring will be white-flowered?
    a) 50%
    b) 100%
    c) 25%
    d) 0%
    e) It depends on whether the traits are linked.

3. Some genes, such as the human ABO blood groups, have more than two alleles. For these genes:
    a) normal dominance relationships are not possible.
    b) a greater proportion of the individuals must be heterozygous than homozygous.
    c) some individuals can be triply heterozygous.
    d) natural selection cannot alter the allele frequencies.
    e) individuals can only possess two alleles.

4. Mendel's Law of Segregation has its physical basis in which of the following phases of the cell cycle?
    a) the orientation of homologous chromosome pairs in metaphase II of meiosis
    b) the orientation of homologous chromosome pairs in metaphase of mitosis
    c) the separation of homologous chromosome pairs in anaphase II of meiosis
    d) the separation of homologous chromosome pairs in anaphase I of meiosis
    e) the orientation of homologous chromosome pairs in metaphase I of meiosis

5. A man with the autosomal recessive disorder phenylketonuria (PKU) and a woman without PKU have a son named Peter, who does not have PKU. Peter is curious about whether his mother is a carrier for PKU. Which of the following facts would allow him to know?
    a) Peter's sister does not have PKU.
    b) Peter's own daughter has PKU.
    c) Peter's maternal grandfather has PKU.
    d) Neither of Peter's maternal grandparents have PKU.
    e) Peter submits his own blood sample to a local genotyping lab, which establishes that he is a carrier for PKU.

6. Mary, who has type O blood, is expecting a child with her husband, who has type B blood. Mary's husband's father has type A blood. What is the probability that the child will have type O blood?
   a) 50%
   b) 25%
   c) 100%
   d) 75%
   e) 0%

7. Assuming that a particular disorder is caused by an allele of a single gene, What feature of a pedigree would allow one to conclude that the disorder was caused by a dominant allele?
   a) Two affected parents have an unaffected child.
   b) Two unaffected parents have an affected child.
   c) An affected mother only has affected sons.
   d) There is no way to tell based on pedigree analysis alone.
   e) All of the descendants of a particular affected person are also affected.

8. Progeria is a genetic disorder that causes numerous symptoms that resemble premature aging in patients. Progeria is caused by a dominant allele, and it always results in fatality before sexual maturity. Which of the following statements must be true about progeria?
   a) Both parents of every progeria patient must also be carriers of progeria.
   b) Progeria must be relatively common.
   c) At least one parent of every progeria patient must also be a carrier of progeria.
   d) Progeria must be more common in some geographic areas than in others.
   e) Every case of progeria must be caused by a de novo mutation.

9. In mice, the allele for brown coat color is completely dominant to the allele for white coat color. Which of the following is true about a true-breeding brown mouse?
   a) If crossed with a white mouse, all the resulting progeny would be brown.
   b) If crossed with a white mouse, and two of the resulting progeny mated, the F2 progeny would be 75% brown and 25% white.
   c) It is homozygous for the brown allele.
   d) If crossed with another brown mouse, all progeny would be brown.
   e) All of the above are correct.

10. Is it possible for a woman to have a X-linked recessive trait; if it is, how can this occur?
   a) No, women cannot have X-linked recessive traits because they are all recessive and women have two X chromosomes.
   b) Yes, a woman will always have an X-linked recessive trait if her father has the trait.
   c) Yes, women can have an X-linked recessive trait if her mother is homozygous for the trait.
   d) Yes, a woman can have an X-linked recessive trait if both her father has the trait and her mother is either homozygous or heterozygous for the trait.
   e) Both b) and c) are correct.

Short Answer

1. The ability to roll your tongue is dominant (R) while the inability to roll your tongue is recessive (r). List the three possible genotypes and the corresponding phenotypes for each.

2. Two parents have the dominant phenotype for tongue rolling yet they produce a child with the recessive phenotype (unable to roll their tongue). What must the genotype of the parents be? What is the chance that they would produce a child with the recessive phenotype (show the Punnett square)?

3. Suppose you have a parent heterozygous for tongue rolling and a parent who is homozygous dominant for tongue rolling. Using the rules of probability, what is the chance that they will have a child who cannot roll their tongue? Show your calculation.

4. Round pea seeds are dominant to wrinkled seeds. If a homozygous round seed plant is crossed with a homozygous wrinkled seed plant, all the offspring will be heterozygous. Suppose you cross two of the heterozygous offspring and produce 400 offspring in the next generation. How many of these would have wrinkled seeds? How many would have round seeds?

5. The inheritance of curly hair illustrates incomplete dominance. When a curly-haired individual reproduces with a straight-haired one, the children all have wavy hair. What offspring would be produced, in what proportions, when two people with wavy hair reproduce? Write your answer and show a Punnett square.

6. Your cousin has three children, all of which are boys. She is convinced that her next pregnancy will produce a daughter since she already has three boys. Based on what you know about probability, is her logic sound? Explain why or why not.

7. A mother is uncertain about the paternity of her newborn baby. Before the DNA analysis is performed, a simple blood test is performed to give some preliminary information. There are two men suspected to be the father in this case so their blood is also collected. Their blood types are as follows:

Mom: Type A   **Suspected father #1**:   Type O
Baby: Type B   **Suspected father #2**:   Type B

- Which man is the most likely father?

- Write an explanation and show Punnett squares to justify your answer.

- To make the situation more complicated, the baby needs a blood transfusion. The mother and both of the suspected fathers offer to donate.

   o   Which of them (write as many as are appropriate) can donate to the baby?

   o   What would happen (specifically) if one of the others (an incompatible match) tried to donate to the baby?

8. Your neighbor has planted a bed of pink snapdragons in his yard. As the growing season progresses, he notices that some white and red snapdragons are popping up in what was supposed to be a bed of pink flowers. Explain to him the reason that this is happening.

Chapter 8
EVOLUTION AND NATURAL SELECTION—DARWIN'S DANGEROUS IDEA

Learning Objectives

- Understand how evolution can be observed in various populations
- Describe Charles Darwin's impact on evolution and the study of biology
- Identify the individuals who influenced Darwin
- Describe Darwin's most important observations
- Explain the four ways evolutionary change can take place
- Identify the difference between evolution and natural selection
- Understand and explain the five different lines of evidence for the occurrence of evolution
- Describe ways evolution can be observed today

Study Guide Outlines

I. Evolution: What is it and what does it look like in action?

- Evolution can be observed by several trials of an experiment with a population of fruit flies.

    o Define the term population.

    o Outline the basic steps of the experiment to increase starvation resistance in fruit flies.

- Define **evolution** with respect to the end result of the fruit fly experiment described in the chapter.

- The findings of the experiment are a result of natural selection. This means the flies in subsequent generations were born with traits or characteristics that allowed them to _____ and _____ better than other flies in the population.

II. Some Background: How the Idea of Evolution and Natural Selection Developed

- While Charles Darwin is the scientist most commonly associated with the study of evolution, several other scientists influenced Darwin and helped lay the groundwork for his ideas.

    o **Curvier** discovered giant fossils that showed no resemblance to living animals. The only explanation for these fossils was:

    o **Lamarck's** studies suggested:

    o **Lyell**, a geologist, presented the idea:

- Charles Darwin's start in the field:

    o Darwin ended up on the HMS Beagle because:

    o Two of Darwin's most important observations include

        1.

        2.

- o   Why did these observations contradict the current scientific thinking of the time?

- Darwin finally published the book titled _____ with all of his ideas and observations.

  - o   Even though Darwin is the sole author of this text, why are both Alfred Russel Wallace and Darwin credited for first describing evolution by natural selection?

III. How can evolution occur?

A. Evolution Occurs When the Allele Frequencies in a Population Change

- In your own words, define evolution.

  - o   The key term *population* should be in the definition above. Scientists can genetically modify mice and an individual can change their physical features, but *individuals* do not evolve.

- Keeping in mind that the allele frequency in a population can change, list the four ways evolutionary change can occur.

  A.

  B.

  C.

  D. Natural Selection
  - o   Natural Selection is not the same as evolution. It is one of the following four agents of change.

B. Mutations

- A mutation is:

- The causes of mutations are the same for both somatic and reproductive cells; however, in considering allele frequencies of a population in studying evolution, we are only concerned with mutations in _____ cells.

- Mutations can be caused by several very different factors. Two nonspecific, common causes involve:

  1.

  2.

       a. List two examples of potential sources of radiation that may cause mutations.

- A mutation can change one allele to a different allele or create a brand new allele. Describe how this can then affect the resulting protein product.

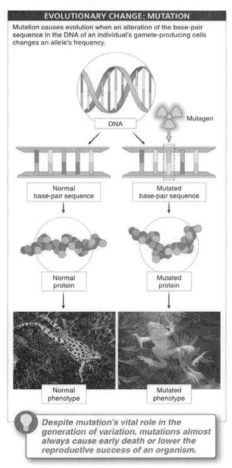

EVOLUTIONARY CHANGE: MUTATION

Mutation causes evolution when an alteration of the base-pair sequence in the DNA of an individual's gamete-producing cells changes an allele's frequency.

DNA

Mutagen

Normal base-pair sequence

Mutated base-pair sequence

Normal protein

Mutated protein

Normal phenotype

Mutated phenotype

Despite mutation's vital role in the generation of variation, mutations almost always cause early death or lower the reproductive success of an organism.

C. Genetic Drift

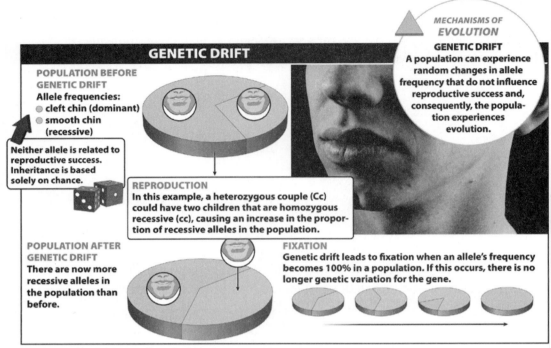

- Genetic drift can be defined as:

- Is the impact of genetic drift seen more greatly in large populations or small populations?

- Genetic drift can occur to the extent where the frequency of an allele in a population is 100%; actually reducing the genetic diversity within the population. This consequence is referred to as _____.

- Genetic drift is different from natural selection because its impact is not directly linked to reproductive success. Give your own example of a trait that might be impacted by genetic drift.

- Two specific impacts of genetic drift are:

  1. Founder effect

     o   Explain the end result of this effect.

  2. Population bottlenecks

     o   Describe what happens in a population bottleneck.

        - These events are due to:

        - How do these events affect evolution?

D. Migration

- Migration is also called _____, which is:

- Can both populations involved in migration experience an impact? Explain.

E. Natural Selection

- This chapter introduces us to evolution through natural selection with the fruit fly starvation resistance experiment. Darwin outlined this agent of change in his famous text. In your own words, define natural selection.

- There are three specific conditions that need to be present in order for natural selection to occur.

    o Condition 1 is _____, which can encompass:

        - physical
        - physiological
        - biochemical
        - behavioral

    Explain what is needed to satisfy this first condition.

    o Condition 2 is _____.

    o Condition 3 is _____. In your own words, summarize this condition:

        - Define sexual selection.

F. A Trait Does Not Decrease in Frequency Simply Because It Is Recessive

- The _____ Law demonstrates that recessive traits do not become less common or less frequent in a population.

- o   The frequency of the dominant allele is represented by a _____

- o   The frequency of the recessive allele is represented by a _____

- o   The genotype represented by p^2 is _____

- o   The genotype represented by q^2 is _____

- o   Explain if the allele frequency of the second generation of kangaroo rats changed in the text's example.

- In order for the Hardy-Weinburg equation to properly predict genotype frequencies, there are two important assumptions of a population that must hold true. What are they?

- If the genotype frequencies observed do not match the predicted frequencies, the population must not be in _____.

IV. Though Natural Selection, Populations of Organisms Can Become Adapted to Their Environment

A. The Misleading Phrase "Survival of the Fittest"

Since **fitness** greatly impacts natural selection and evolution, and it doesn't refer to how fast someone can run the mile, define fitness in respect to natural selection.

In two sentences or less, define the three aspects important to the (evolutionary) fitness of an individual.

1.

2.

3.

B. Adaptation

Adaptation refers to _____
and _____ .

C. Artificial Selection

The three conditions necessary for natural selection are also satisfied in artificial selection, but the difference is that reproductive success is determined by _____ rather than nature.

Explain a situation where a breeder or farmer would benefit from utilizing natural selection.

D. Natural Selection

1. Natural selection can lead to change, but not "perfection."

    In you own words, briefly explain the factors that support this statement: "A population will never be, what some might consider, perfect."

2. Natural selection can select for change in simple traits.

    Many populations have a range of traits that can be greatly influenced by natural selection. The variation of traits results in a variation of phenotypes. Complete the following chart to highlight the influence of natural selection.

Changes can occur in three major ways:

Type of selection	What occurs?	Example

3. Natural selection can select for change in complex traits.

Some traits that are more complex, such as certain behaviors, may also be affected by natural selection. The three conditions necessary for natural selection to occur are still satisfied, even though some of the traits involve multiple physiological systems.

Give an example of how natural selection could affect a specific behavior.

V. Understanding the Evidence Involved in Evolutionary Change

The following five unique areas of evidence help us better understand Darwin's original thoughts and have advanced all areas of biology.

A. Fossil Record

• While fossils are often though of as "old bones," technically, fossils are:

• Briefly explain how radioactive isotopes are utilized in fossil records.

- List three ways the analysis of fossil records helps to provide evidence for the process of natural selection.

    1.

    2.

    3.

B. Biogeography

- Define biogeography in your own words.

- What is unique about the biogeographic pattern of some of the organisms in Australia?

C. Embryology and Anatomy

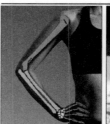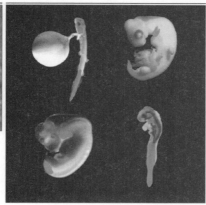

- By examining the vertebrate embryo and various aspects of anatomy, one can identify common features, or _____.

- Briefly explain what a vestigial structure is and list a common example.

D. Our DNA

- New technology has allowed us to sequence, or map out, our genetic code (DNA). This has also been successful in many other species.
- Patterns can be noted in examining the DNA sequence of various organisms. If you were examining the DNA of a brother and sister, would the sequences look more or less similar than the DNA of the brother as compared to his cousin?

   o Why?

- To compare DNA sequences between species, an individual gene that produces a specific protein can be examined. Proteins are made up of building blocks called amino acids. Therefore, comparing the amino acids will help compare the DNA sequences.

- In examining the beta chain of the hemoglobin protein, it has been found that humans have _____ (number of) amino acids in the beta chain.

- In comparison:

    o Rhesus monkeys have _____ (number of) amino acids that are
      different as compared to the human sequence.
    o Dogs have _____ amino acids that are different.
    o Birds have _____ amino acids that are different.
    o Lamprey eels have _____ amino acids that are different.

- Explain what this means:

E. Using the Scientific Method to See Evolution in Action

- Just as the fruit fly starvation resistance experiment illustrated in the first section
  of the chapter, carefully planned and executed experiments both in the lab and in
  the field allow us to observe and learn about evolution as it occurs.

- The evolution of bacteria has led to a very important public health concern of
  today.
    o Define an "antibiotic-resistant" strain of bacteria.

    o How have bacteria become resistant to pharmaceuticals such as
      penicillin?

Testing and Applying Your Understanding

Multiple Choice (For more multiple choice questions, visit www.prep-u.com.)

1. Evolution is defined as:
    a) a change in the frequency of alleles in a population over time.
    b) a change in the frequency of a morphological trait in a population over time.
    c) a progressive "ladder" of changes from most primitive organisms to most advanced organisms.
    d) a change in a morphological trait of an individual during its lifetime.
    e) survival of the fittest.

2. To demonstrate evolution by natural selection, all of the following conditions must be satisfied EXCEPT:
    a) variation for a trait.
    b) heritability of a trait.
    c) differential reproductive success.
    d) genetic drift.
    e) All of the above are necessary for evolution by natural selection.

3. _____ selection favors organisms that have character values at both extremes of the phenotypic distribution.
    a) Disruptive
    b) Stabilizing
    c) Directional
    d) Intense
    e) Intermittent

4. Which of the following statements is NOT consistent with evolution by natural selection?
    a) Individuals in a population exhibit variation, some of which can be inherited by their offspring.
    b) Individuals change during their lifespans to fit their environment better, and these changes can be inherited by their offspring.
    c) Natural selection can lead to speciation.
    d) Individuals that reproduce most successfully are more likely to have offspring that also reproduce successfully if the environment remains stable.
    e) Certain individuals in a population have a higher rate of reproductive success than other individuals due to a variety of environmental and developmental factors.

5. All of the following statements are true about mutations EXCEPT:
   a) The origin of genetic variation is mutation.
   b) A mutation is any change in an organism's DNA.
   c) Mutations are almost always random with respect to the needs of the organism.
   d) Most mutations are beneficial or neutral to the organism in which they occur.
   e) The mutation rate can be affected by exposure to radiation.

6. A population is:
   a) a group of species that share the same habitat.
   b) a group of individuals of the same species that have the potential to interbreed.
   c) a group of individuals of the same species that live in the same general location and have the potential to interbreed.
   d) a group of individuals of related species that live in the same general location and have the potential to interbreed.
   e) a group of individuals of the same species that live in the same general location and have the same genotypes.

7. Birth weight in human babies is generally:
   a) subject to stabilizing selection.
   b) not a heritable trait.
   c) a strong predictor of adult weight.
   d) subject to disruptive selection.
   e) All of the above.

8. In a fish population in a shallow stream, the genotypic frequency of yellowish-brown fish and greenish-brown fish changed significantly after a flash-flood randomly swept away individuals from that stream. This change in genotypic frequency is most likely attributable to:
   a) gene flow.
   b) disruptive selection.
   c) directional selection.
   d) genetic drift.
   e) the founder effect.

9. Which of the following best explains why genetic bottlenecks and founder effects are evolutionarily important?
   a) In both cases, strong selective pressures lead to fast directional selection.
   b) Both result in stabilizing selection due to strong selective pressures.
   c) Both result in small populations subject to genetic drift.
   d) Both result in increased fitness.
   e) Both a) and d) are correct.

10. The phenotypic trait of polydactyly, where an individual has extra fingers or toes, is one symptom of Ellis-van Creveld syndrome. This syndrome is more commonly found in Old Order Amish populations. Which of the following is a possible explanation for why this occurs within this population?
    a) The Old Order Amish experience the "founder effect," where all the members of a population descend from a small group of founding individuals.
    b) The Old Order Amish live near biowaste dumping sites that increase their number of genetic mutations and result in strange disorders.
    c) The Old Order Amish experience the "bottleneck effect" because they are physically isolated from other communities.
    d) Both answers a) and c) are correct.
    e) None of the above is correct.

Short Answer

1. Explain the difference between evolution and natural selection.

2. Does evolution occur in individuals or populations? Explain.

3. In describing the results of evolution (e.g. "The finches evolved to have harder beaks"), it often seems as if populations changed with intent or to reach a specific goal. Explain why this does not accurately describe how evolution occurs.

4. What is the difference between natural selection and genetic drift?

5. Why would genetic drift be important to endangered species?

6. Using the fruit fly experiment explained in the text, explain how the three conditions necessary for natural selection were in place.

7. Certain crustaceans have exoskeletons (shells) for protection. Some types of exoskeletons are thinner allowing for the organism to move quickly out of harm's way. However, the thinner exoskeletons are more easily penetrated or punctured by a predator. The evolution of the exoskeletons' thickness in certain populations would reflect what manner of natural selection?

8. Explain how bats that have developed superior echolocation are an illustration of evolutionary adaptation. Be specific.

9. An organism itself isn't necessarily "fit" to survive, but do the alleles "survive" in a population? Explain why or why not.

10. Give an example that demonstrates each of the five lines of evidence for evolution.

11. What is a vestigial structure? Include an example in your explanation.

12. Could the recent development of antibiotic-resistant bacteria be a result of natural selection, artificial selection, or both? Justify your answer.

Chapter 9
EVOLUTION AND BEHAVIOR—COMMUNICATION, COOPERATION, AND CONFLICT IN THE ANIMAL WORLD

Learning Objectives

- Understand that the behaviors displayed by animals are shaped by natural selection and are performed in an attempt to increase fitness
- Compare and contrast learned and innate behaviors
- Discuss how kin selection and reciprocal altruism lead to behaviors that appear to be completely altruistic
- Explain how altruistic behaviors can be predicted
- Characterize the factors that contribute to an individual's inclusive fitness
- Explain why selfish genes tend to increase in a population
- Describe the most common types of communication signals among animals
- Compare and contrast the differences between paternal investments between genders
- Explain how parental uncertainty can shape reproductive behavior
- Describe how discrimination and competition in reproductive behaviors can be predicted
- Explain the four types of female choosiness in selecting mates
- Discuss how mate guarding can help decrease paternal uncertainty and increase paternal investment
- Compare and contrast monogamous and polygamous mating systems
- Explain how sexual dimorphism in animals can be an indicator of parental investment

Chapter Outline

I. Behaviors Are Traits That Can Evolve

- The influence of natural selection is very apparent when studying the behaviors of animals. Animals perform certain behaviors, even ones that might seem quite ridiculous to an observer, in a quest to increase their fitness.

    o The foods that animals select are an example of this. Why is it that animals tend to prefer calorie-rich foods when given a choice?

- Describe the term **behavior**:

- List three examples of various types of animal behaviors:

A. Innate Behaviors

- While genetics plays a strong role in the behaviors of animals, there is an influence that comes from the _____ as well.

- Behaviors are shaped by a variety of influences. Those that are completely genetic and require no learning are referred to as _____ or innate behaviors. Animals perform these behaviors the same way over their lifespan.

- An example of a completely innate behavior that can be performed by any animal, does not change over time, and is always completed once initiated is called a

  _____.

  o   An example of this sort of behavior would be:

B. Learned Behaviors

- Learned behaviors require various inputs from the environment and can be adjusted over the lifespan of the animal. A behavior that is performed by nearly all animals and is learned easily is considered to be shaped by _____ learning.

  o   Explain why certain behaviors (such as fear of snakes by human babies) seem to be easier to learn than others (such as fear of guns by human babies).

C. Complex-Appearing Behaviors Don't Necessarily Require Complex Thought

- Animals respond to specific cues in their environment. While the animal responding to a cue might appear to be very innovative, they are really performing a specific behavior in response to a simple environmental cue.

- This allows the animal to perform behaviors without having to consciously attempt to increase their reproductive success. However, the behaviors themselves ensure increased reproductive success.

    o   Explain how egg retrieval behavior in geese exemplifies this principle.

II. Cooperation, Selfishness, and Altruism Can Be Better Understood With an Evolutionary Approach

- Behaviors that cost one individual while benefitting another individual are called _____ behaviors. While many behaviors may appear to benefit one individual while costing another, there are very few behaviors that truly involve a cost to an organism with no benefits associated with that cost.

- While behaviors may appear altruistic, natural selection has really produced behaviors that are more _____ in nature.

A. Kin Selection

- **Kin selection** is one factor that leads toward apparent altruistic acts. In this case, an individual is performing a seemingly selfless act because the act has the ability to increase the fitness of that individual's _____, which can compensate for the individual's own reduced fitness.

- Animals are not actually able to determine the relatedness of animals around them. How do they determine when to perform altruistic behaviors?

- Belding's ground squirrels are known to make alarm calls to warn other squirrels of danger, even when making the call may lead to death. Explain why females are more likely to display these acts than males.

- The **inclusive fitness** of an individual is a combination of:

    1. an individual's reproductive output or _____ fitness, and

2. the reproductive output of individuals who are close relatives to the animal performing the altruistic act, termed _____ fitness.

B. Reciprocal Altruism

- Some animals display cooperative behaviors that assist other animals that are not their relatives. By doing so, the animal performing the altruistic behaviors pays a cost. Explain why an animal would ever perform this sort of behavior for a non-relative.

- Describe the conditions that must be met in order for **reciprocal altruism** to evolve.

    1.

    2.

    3.

C. Fitness Is Relative to Your Environment

- Sometimes animals display behaviors that appear to reduce their fitness. While this seems to contradict what we might expect, there are reasons that these sorts of behaviors persist. While our instincts may not have changed, our environments have.

- Using humans as an example, offer an explanation as to why we might feel compelled to help other humans that we may never meet.

D. Group Selection

- The concept of **the selfish gene** is apparent in many behaviors performed by animals. Selfish genes increase when they benefit the individual carrying them even if this comes at the expense of the group.

  o   Provide an example of a selfish gene:

- In very rare cases, _____ selection may occur in which behaviors evolve that increase the good of the population at the expense of the individual animal.

III. Sexual Conflict Can Result from Disparities in Reproductive Investment by Males and Females

- While most of us have our own ideas about the features that differentiate between males and females, the true difference comes down to the size of the gametes. In a relative comparison of size, the **female** gamete is _____ , whereas the **male** gamete is _____.

- In addition to size, there are some other features of female gametes that differentiate them from male gametes. Those differences include:

A. Parental Investment

- The maximum reproductive output for males is always higher than that of females. Explain why this is the case.

- In addition to differences in gametes and energetic investment, females also have a greater parental investment during gestation. Provide several examples of this.

- Parental investment can sometimes be equalized in certain species. Explain how each of the following can help with this equalization of parental investment:

    o   external gestation

    o   lack of lactation

    o   external fertilization

- Reproductive behavior in male mammals can be influenced by the possibility of a female producing offspring that are not the male's progeny. This is termed
_____.

B. Maximizing Reproductive Success

- In looking at sex-related behaviors, the gender that invests less will be
_____ (more or less) discriminating and will be
_____ (more or less) competitive for mates.

- Females that have a high level of investment in reproduction display a high level of choosiness about their mates. There are four types of choosiness that females use in order to select males with quality genes and plentiful resources. Explain each of these four types and provide an example for each.

    1.

    2.

    3.

    4.

- Males typically display aggressive fighting behaviors more so than females. Explain why this is the case.

- Males that invest significantly in parental care have a vested interest in making sure that the offspring produced are theirs. This often leads to **mate guarding**. What purpose does this behavior serve?

- Mate guarding can be achieved in several ways including extensive periods of copulation or attempts to seal the female's reproductive tract. Provide several examples of this type of behavior.

C. Parental Care and Mating Systems

- Two major categories of **mating systems** are **polygamy** and **monogamy**. Describe how these two systems of mating differ from each other.

- Within polygamous mating systems, there are two subcategories. When individual males mate with multiple females, this is called _____. When individual females mate with multiple males, this is referred to as

  _____.

- Pair bonding occurs when a male and female spend a large percentage of their time with each other. This gives the impression that they are monogamous. However, this may not be the case. Explain why.

- The type of mating system used by a population can often be predicted by looking at the differences in the size and appearance of males and females of the population. These differences are referred to as _____.

IV. Communication and the Design of Signals Evolve

- Animals have many reasons for needing to be able to communicate with each other. Describe the three most common types of **communication** among animals:

1. Chemical:

2. Acoustical:

3. Visual:

- The **waggle dance** is an example of an elaborate communication behavior performed by honeybees. What is the purpose of this behavior?

- How does **language** differ from other forms of animal communication?

- Forms of communication signals that cannot be faked by animals provide the most accurate information to other animals. These are referred to as _____ signals.

Testing and Applying Your Understanding

Multiple Choice (For more multiple choice questions, visit www.prep-u.com.)

1. Exclusive male parental care is much more prevalent in fish than mammals because:
    a) most fish reproduce by external fertilization, whereas most mammals do not.
    b) fish are less likely to live in habitats with high-quality resources.
    c) fish are less intelligent and so do not develop pair bonds.
    d) fish do not have the option of raising young in a defended nest site or den.
    e) both male and female fish lactate.

2. Which of the following is NOT an important condition for the maintenance of reciprocal altruism?
    a) relatedness to the individual being helped
    b) lots of opportunities for helping others and being helped
    c) the ability to recognize specific individuals
    d) repeated interactions with the same individuals
    e) All of the above are important in the maintenance of reciprocal altruism.

3. Kin selection is defined as:
    a) selection for a behavior that lowers an individual's own chances of survival or reproduction, but raises those of a relative.
    b) selection for a behavior that raises an individual's own chances of survival or reproduction, as well as those of a related individual.
    c) selection for a behavior that lowers an individual's own chances of survival or reproduction, but raises those of another member of the population who may perform altruistic acts for the benefit of that individual.
    d) selection that operates not just on one individual, but on all of its relatives as well.
    e) None of the above is correct.

4. When a goose spots an egg outside of its nest, the goose gets out of the nest and rolls the egg back. Once started, a goose continues the egg-retrieval movement all the way back to the nest, even if the egg is taken away during the process. This is called:
    a) a learned behavior.
    b) supernormal stimulus.
    c) maximizing inclusive fitness.
    d) a fixed action pattern.
    e) prepared learning.

5. From an evolutionary perspective, behavior can be viewed best as:
    a) part of a phenotype.
    b) a trait subject to drift and mutation, but not natural selection.
    c) a trait that arises by learning and not natural selection.
    d) non-heritable.
    e) All of the above are correct.

6. Of the following social acts, which is least likely to evolve?
    a) a spiteful act
    b) a selfish act
    c) a cooperative act
    d) an altruistic act
    e) All acts have an equally likely chance of evolving.

7. Sexual dimorphism is a good predictor of mating systems because:
    a) when males are bigger than females, males are able to contribute more to their offspring than females resulting in polyandry; therefore, sexual dimorphism determines mating system.
    b) mating systems cause variance in reproductive success, which results in sexual dimorphism; therefore, when one is known the other can be inferred.
    c) the degree of sexual dimorphism and the type of mating system present in a species are both determined by the difference in parental investment between the male and female; therefore, when one is known the other can be inferred.
    d) sexual dimorphism causes a difference in parental investment between males and females which results in different variance in reproductive success; therefore, when one is known the other can be inferred.
    e) only polygynous mating systems exhibit sexual dimorphism; so if sexual dimorphism is present, the mating system is known.

8. In mammals, as well as many other species, males generally compete for females. The best explanation for this phenomenon is:
    a) females have higher fitness.
    b) females are choosy.
    c) females have higher parental investment.
    d) males are more aggressive.
    e) females are better looking.

9. Which of the following best illustrates an instinctive behavior in cats?
    a) scratching at the door to get in or out
    b) knowing when it is time to get fed
    c) being afraid of dogs
    d) hunting and killing
    e) All of the above are instinctive behaviors in cats.

Short Answer

1. While the fancy plumage of peacocks might seem very burdensome to the animal, this adaptation persists throughout generations. Why is this the case?

2. Explain why is it easier for animals to learn certain behaviors than others.

3. Reproductive success is key to fitness. From an evolutionary perspective, why would it be that some animals fail to reproduce while assisting other animals in rearing their offspring?

4. Why are most workers in a honeybee colony female?

5. Why is reciprocal altruism much more common in humans as compared to other animal species?

6. Suppose you have been working on training your new pet dog. While she seems to learn certain cues very quickly, there are others that she just doesn't seem to respond to easily. Based on what you know about learning, what might account for the difference in your new dog's ability to learn new cues?

7. How can you determine whether a male or female of a particular species will be more discriminating in their reproductive behaviors? Which gender will compete more for mates?

8. Why might a male that provides an appealing nuptial gift be more likely to mate with a female than a male that does not present a nuptial gift?

9. From an evolutionary perspective, while it may seem that males should mate with as many females as possible, sometime they sacrifice additional matings in order to provide a paternal investment. Why does this sort of behavior actually increase the fitness of a male more so than attempting to mate with additional females?

Chapter 10
THE ORIGIN AND DIVERSIFICATION OF LIFE ON EARTH—UNDERSTANDING BIODIVERSITY

Learning Objectives

- Define life
- Outline the conditions and evidence that support how life on earth was formed
- Explain how to identify and name a species
- Compare and contrast the biological species concept and the morphological species concept
- Understand the various processes of speciation
- Understand the purpose of a phylogenic tree and what it can demonstrate
- Define the difference between analogous traits and homologous features
- Compare and contrast microevolution and macroevolution
- Explain how adaptive radiation and extinction impacts evolution
- Understand the current biodiversity found in the three domains

Chapter Outline

I. Life on Earth Most Likely Originated from Non-living Materials

- In order to explore the question "How did life begin?" the term **life** must first be defined. Provide the two important aspects or conditions needed for life.

  1.

  2.

- Early forms of life, such as bacteria-like cells, seem nothing like the life forms that inhabit the earth today. In this context, define **biodiversity**.

- While there are different suggestions as to how the first life forms arose, much of the evidence points to life originating on earth in several phases.

A. Phase 1

- Describe the content of the earth's atmosphere around the time life originated and how it was different from today's atmosphere.

- The chemical reactions necessary to support life require small molecules that contain carbon and hydrogen. Outline the experiment that demonstrates how these molecules were likely formed during the origin of life.

    o   What did the experimental results reveal?

B. Phase 2

- Without the ability to replicate, life cannot be supported. Therefore, Phase 2 explores how the carbon- and hydrogen-containing molecules can be used as building blocks for more complex molecules.

- Explain the **RNA world hypothesis**.

- What characteristic or condition of life is *not* supported by Phase 2?

C. Phase 3

- Explain why a membrane is relevant and important to the third phase of the development of life.

- Define **microspheres**.

II. Species Are the Basic Units of Biodiversity

A. What is a species?

- With the diversity of living organisms on earth it is helpful to group like organisms together. _____ is the term used to describe organisms that are members of the same group.

- What are the key characteristics of the **biological species concept**?

   o   This concept does not rely on appearance or looks of organisms, but rather relies on **reproductive isolation** to distinguish between groups or species. What does reproductive isolation mean?

      ▪   If two organisms are physically separated, does that mean they are of two different species? Explain.

      ▪   Why is the term "natural conditions" an important part of this concept?

- Explain the following barriers to reproduction. Include an example.
   o   **Prezygotic barriers**

   o   **Postzygotic barriers**

- It is important to note that while the biological species concept works very well for plants and animals, it is not possible to apply this concept to all living organisms.

B. Naming Species

- With the immense diversity found on earth, a robust system is needed to organize different species. Who developed the system that is still used today to classify organisms?

- How does the system work?

○ Define and give an example of a **genus.**

○ Define and give an example of a **specific epithet**.

- List the levels of organization or classification from more specific to more broad.
  1. genera
  2.
  3.
  4.
  5.
  6.
  7. **domain**
     ○ List the three domains.

- What are common names of plants and animals often based upon?

C. Difficulties in Identifying and Classifying Species

- Highlight the instances where the biological species concept is inadequate in classifying organisms. List the difficulty and provide an example.
  1.

  2.

  3.

  4.

  5.

- How, specifically, does the **morphological species concept** differ from the biological species concept?

D. How do new species arise?

- The number of species thought to inhabit the earth ranges from _____ to

  _____.

- The term for the process by which new species arise is _____.
    - This process requires two phases. Briefly describe:
        - Reproductive isolation

        - Genetic divergence

- While the first phase of reproductive isolation often requires two populations to be in separate geographical areas, it is not absolutely necessary.
    - Speciation with geographic isolation is called _____.
    - Speciation without geographic isolation is called _____.

- Give an example of **allopatric speciation** and describe how the two species came to be.

- **Sympatric speciation** is most common in what type of organisms?
    - Explain the two ways reproductive isolation can occur if geographic isolation is not a factor.

      1.

      2.

- What process is more likely to take longer, reproductive isolation or genetic divergence? Explain.

III. Evolutionary Trees Help Us Conceptualize and Categorize Biodiversity

A. Tree of Life

- What is the purpose of a phylogenic tree? Include a definition of **phylogeny** in your answer.

- After a **speciation event**, what happens to the tree?

- What is the **node** of the tree?

- Can an observer determine how advanced or primitive a group is, or how closely groups are related, by examining an evolutionary tree? Explain.

  o   What is a **monophyletic** group?

- How has the use of DNA sequences changed the construction of evolutionary trees?

B. Common Structures versus Common Ancestry

- Basing evolutionary relationships solely on physical features can be misleading. Give an example of how the introduction of DNA sequences has changed the mapping of phylogentic trees.

- How are **analogous** traits and **homologous** features different?

- What is **convergent evolution**?

IV. Macroevolution Gives Rise to Great Diversity

A. Macroevolution versus Microevolution

- Evolution is a change in a population's _____ _____ over time.

- The scale of evolution can vary when comparing the length of time allowed for changes to take effect.
    - **Microevolution** refers to:

    - **Macroevolution** refers to:

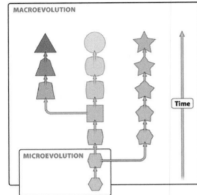

B. Pace of Evolution

- There are examples of evolutionary change occurring at different paces over time.
    - Define gradual change and give an example.

    - Define **punctuated equilibrium** and give an example.

- Explain the two misconceptions of punctuated equilibrium that tend to undermine the understanding of evolutionary theory.

    1.

    2.

C. Adaptive Radiations

- About 65 million years ago the earth's environment experienced a catastrophe that led to a drastic and rapid change in the species that then roamed the earth.

- Define **adaptive radiation.**

- List, define, and give an example for each of the three different events that can trigger adaptive radiation.

    1.

    2.

    3.

D. Extinctions

- What is an **extinction**?

- The two different categories of extinctions differ in their impact and what may trigger the event.

    o   What happens during a **background extinction**?

        ▪   What is its cause?

    o   What happens during a **mass extinction**?

        ▪   What is its cause?

        ▪   How many mass extinctions have occurred in the past 500 million years? Describe what is considered the worth catastrophe in earth's history—the "Great Dying."

V.   An Overview of the Diversity of Life on Earth: Organisms Are Divided into Three Domains

A. The History of Biological Domains

- Linnaeus classified all living organisms into the three kingdoms of animal, plant, and mineral. The discovery of microscopic organisms called _____ started to change the classification system.

- A five-kingdom system then emerged. This newer system separated organisms based on what?

    o List the five kingdoms and describe the cell type(s) in each.

- Carl Woese helped change the five-kingdom system by examining and comparing the _____ of living organisms, a genetic informational molecule found in all living organisms.

- The information Woese gathered, along with the discovery of archaea, lead to the creation of domains.
    o List the three domains.

    o Where did Woese locate the domains in Linnaeus' original classification system?

- Even though the three-domain, five-kingdom system is widely accepted, there will always be exceptions.
    o What is **horizontal gene transfer** in bacteria and how would this complicate the classification of an organism?

- With the current tree of life, list the events that occurred from the origin of the first organic molecules.

    1.

    2.

    3.

- List some characteristics of viruses that could classify them as living, and list some characteristics that would classify them as non-living.

B. Bacteria

- Describe some of the environments in which bacteria are found.

- Describe the structure of bacteria.

- Are all bacteria "bad"? Give some examples of the various functions of bacteria.

C. Archaea

- List the similarities and differences between archaea and bacteria.

- Archaea are often thought of as "extreme" for the environments in which they are found. Characterize the five different groups of archaea.

D. Eukarya

- Despite the diversity of the organisms found in the domain eukarya, list the characteristics they all have in common.

- Complete the chart by listing each of the domain's kingdoms and provide some examples.

Kingdom	Example/Description

Testing and Applying Your Understanding

Multiple Choice (For more multiple choice questions, visit www.prep-u.com.)

1. The classic "Miller experiments," performed in the 1950s by Harold Urey and Stanley Miller, were the first to show that:
   a) cosmic microwave background radiation (CMB), a type of electromagnetic radiation, fills the entire universe and provides support for the "big bang" model of the origin of the universe.
   b) naturally occurring antibiotics, such as penicillin, could be used to treat bacterial diseases.
   c) simple organic molecules, such as amino acids, could form spontaneously in the laboratory under chemical conditions mimicking those of primitive earth.
   d) the average bill size of a population of Darwin's finches changed with environmental conditions, proving that microevolution can occur within a human lifetime.
   e) water (H_2O) could be separated into atmospheric oxygen and H_2 gas.

2. Which of the following molecules was NOT present in the pre-biotic environment?
   a) molecular dioxygen (O_2)
   b) hydrogen sulfide (H_2S)
   c) methane (CH_4)
   d) water (H_2O)
   e) ammonia (NH_3)

3. A phylogenetic tree:
   a) can illustrate selection pressures that have acted on taxa.
   b) is a hypothesis about evolutionary relationships.
   c) shows the similarity between different taxa.
   d) must be rooted to properly convey evolutionary relationships.
   e) shows the number of character changes that have occurred between taxa.

4. For a new species to arise:
   a) a barrier to gene flow between two or more populations must exist.
   b) a vacariant event must occur.
   c) the parent species must become extinct.
   d) post-zygotic barriers must first be established.
   e) a population must be physically and reproductively isolated from others of its species.

5. The Biological Species Concept:
   a) states that a species is any group that shares similar DNA.
   b) is the best and most accurate species concept.
   c) is infrequently used because it is inapplicable to many organisms.
   d) states that any organisms that are biologically similar constitute a species.
   e) defines a species as any group of actually or potentially interbreeding species that produce viable offspring.

6. You are a taxonomist and your job is to classify and name newly discovered organisms. On what do you base your criteria in order to form your classifications?
    a) Their evolutionary relationship to existing known organisms.
    b) Their similar appearance to existing known organisms.
    c) Their proximity to existing known organisms.
    d) Their similar metabolism methods to existing known organisms.
    e) Their ability to reproduce with existing known organisms.

7. The idea of "punctuated equilibrium" suggests that species will show little to no evolutionary change throughout their history. When evolution does occur, according to this idea, it happens sporadically and relatively quickly compared to the species' full duration on earth. This idea of evolution in spurts challenges which of the components of Darwin's theory of evolution?
    a) steady change
    b) gradualism
    c) multiplication of species
    d) Both a) and b) are correct.
    e) None of the above.

8. Sequencing DNA from different organisms has been a breakthrough for phylogenetics because:
    a) it has reduced the impact of antibiotic resistance in bacteria.
    b) we now have a method to determine the relatedness of organisms that leave no fossils.
    c) we can now build heat-tolerant animals.
    d) we can now recreate genomes from the past.
    e) All of the above.

9. Horses and donkeys can breed and produce sterile offspring known as mules. Horses and donkeys remain separate species because of this hybrid sterility, which is:
    a) an allopatric barrier to reproduction.
    b) a sympatric barrier to reproduction.
    c) a postzygotic barrier to reproduction.
    d) a prezygotic barrier to reproduction.
    e) a good thing.

10. *Homo sapiens* is the name of a species. *Homo* is the name of a genus. Hominidae is the name of a _____. Primate is the name of an order. Mammal is the name of a class. Animal is the name of a kingdom. Eukaryote is the name of a domain.
    a) subspecies
    b) genera
    c) family
    d) variety
    e) grade

Short Answer

1. If a semi-permeable barrier, such as a phospholipid bilayer, did not exist how would this affect the origin of life and our definition of life?

2. Bowerbirds are very unique birds who reside in parts of Australia. Part of their complex mating ritual involves males building bowers, or a type of nest, to attract the female. The Vogelkop bowerbirds build teepee-like nests with bright objects and flowers. The Satin bowerbirds build a very different structure containing two walls of twigs and sticks adorned with blue and green objects. Females will approach and inspect the bower, but only certain females are attracted to particular bowers. Explain whether the two groups would be considered separate species or the same species according to the biological species concept. Be specific.

3. During a severe drought, a waterway connecting two bodies of water dried up. This left a separate larger lake and a smaller pond. A species of fish that once populated the entire area was reduced in number. After some time, an evaluation of both bodies of water showed the fish in each of the separate bodies were vastly different. Describe the type of speciation event that occurred.

4. Does adaptive radiation fit into the idea of punctuated equilibrium? Explain why or why not.

5. The Giant Anteater is found in South America and can weigh upwards of 80 lbs. The Spiny Anteater who resides in Australia only weighs up to about 20 lbs. Although they differ in many ways, both animals have long, thin, sticky tongues to forage for ants and termites. Would this trait or feature be considered homologous or analogous? Explain and support your answer.

6. Complete the following chart regarding the three domains.

Characteristics	Domains		
	Bacteria	Archaea	Eukarya
Unicellular or multicellular?			
Cellular structure			
Give an example of how the domain is unique from the other two domains.			

7. Despite all of the diversity between the three major domains, describe a similar characteristic or need shared by organisms in all three domains.

Chapter 11
ANIMAL DIVERSIFICATION: VISIBILITY IN MOTION

Learning Objectives

- Describe the three characteristics that define all animals
- Describe the criteria used to classify animals into various phyla
- Explain the features that define an animal's evolutionary success
- Compare and contrast the features of vertebrates and invertebrates
- Describe example species and adaptations of each major animal phylum
- Explain the unique adaptations seen within the invertebrate groups
- Discuss the unique characteristics that have led to the evolutionary success of the arthropods
- Describe the four adaptations needed by terrestrial vertebrates
- Discuss the characteristic structures associated with the chordates
- Discuss the evolution of primates including humans

Chapter Outline

I. Animals Are Just One Branch of the Eukarya Domain

- There are several characteristics that define all animals. The three characteristics that are easiest to identify are:

  1.

  2.

  3.

- Because of the diversity amongst animals, there are four key questions that are asked in order to classify animals into groups. Those questions include the following:

  1. *Are defined tissues with specialized cells present?*

     o   An example of an animal without tissues would be:

     o   An example of an animal with defined tissues would be:

2. *What type of symmetry does the animal have?*

    o   **Radial symmetry** is defined as:

    o   **Bilateral symmetry** is defined as:

3. *How does the animal's gut develop?* Describe the differences between:

    o   **Protostomes**:

    o   **Deuterostomes**:

4. *Does the animal molt or have a growing skeleton?* This applies to animals with
    _____ symmetry that are also _____.

    o   Explain the difference between a molting and a growing skeleton.

- In terms of evolutionary success, animals must simply survive. The three things animals
  need to do in order to stay extant include:

  1.

  2.

  3.

- All members of the animal kingdom are organized into _____ major phyla.

II. Invertebrates Are Animals without a Backbone

- The invertebrates have the largest amount of diversity within the animal kingdom. Among the invertebrates, most are protostomes. However, there is one group within the invertebrates that contains dueterostomes. This is the _____ group, and several examples of these animals would be:

III. Across Several Evolutionary Transitions, the Invertebrate Animals Diversified

A. Sponges

- Sponges lack some features that one might expect from a "typical" animal. Describe the two characteristics of sponges that make them particularly different from other groups of animals.

    1.

    2.

- Describe how the body plan of the sponge makes it particularly well-suited to feeding.

- Sponges have some unique reproductive adaptations that allow them to reproduce sexually or asexually. Explain the characteristics of each sort of reproduction,

    o   Asexual reproduction:

    o   Sexual reproduction:

B. Cnidarians

- The cnidarians (jellyfish, corals, and sea anemones) are noted for their defined tissues and _____ symmetry. Describe the characteristics of each major group.

    o   Coral:

    o   Sea anemones:

    o   Jellyfishes:

- There are two major body forms seen among the cnidarians—the polyp and medusa forms. Describe each form.

    o   Polyp:

    o   Medusa:

C. Worms

- Worms exist in _____ of the nine major phyla. Characteristics of the three major groups of worms are as follows:

    o   Flatworms:

    o   Roundworms:

    o   Segmented worms:

D. Molluscs

- There are many adaptations amongst the molluscs, and there are three features typical to most species in this group. Those three features include:

  1.

  2.

  3.

- The three major groups of molluscs share many of the same features but have different body plans. Describe the characteristics of each group.

  o Gastropods:

  o Bivalves:

  o Cephalopods:

E. Insects

- Insects outnumber all other forms of life in species diversity. Some groups of insects have many more species than all animal groups combined.

- All of the arthropods, including insects, are protostomes. They have some unique adaptations including:

  o Segmentation:

  o Exoskeletons:

  o Jointed legs:

- The lifecycle of insects has three unique stages that the animals progress through in their **metamorphosis**. Describe the major events of each stage.

  1. Larval form:

  2. Pupal form:

  3. Adult form:

- Most insects have a complete metamorphosis, but some do not. Using examples, contrast the difference between complete and incomplete metamorphosis.

F. Other Arthropods

- In addition to insects, there are three other major groups of arthropods. For each of these groups, indicate the unique characteristics.

  o   Millipedes and centipedes:

  o   Arachnids:

  o   Crustaceans:

G. Echinoderms

- The echinoderms are very unique in that some of the characteristics seen in the development of these animals are different from the characteristics displayed by the adult animal. One example of such a change involves symmetry. During development, the type of symmetry displayed is _____, but in the adult form, _____ symmetry is observed.

- Some examples of echinoderms include:

- Describe several additional characteristics of echinoderms.

IV. The Phylum Chordata Includes Vertebrates, Animals with a Backbone

- All vertebrates are members of the phylum _____. There are several distinct body features that distinguish members of this phylum from other animals. Describe each of these structures.

  o **Notochord:**

  o **Dorsal hollow nerve cord:**

  o **Pharyngeal slits:**

  o **Post-anal tail:**

- The phylum **Chordata** can be further divided into three sub-phlya whose formal names are: _____, _____, and _____. Characteristics of each sub-phylum include:

  o Tunicates:

  o Lancets:

  o Vertebrates:

A. The Evolution of Jaws and Fins

- Vertebrates eat other organisms and this action requires a mouth. The earliest vertebrates had to swim to find their prey and they had mouths but were lacking jaws. Because obtaining food required swimming, there was a co-evolution of jaw and fin development.

  o The two groups of **jawless fishes** that are in existence today include the _____ and the _____. Describe each of these groups.

    1.

    2.

  o What purpose do fins serve in helping an organism get to its prey?

- There are three major groups of jawed fishes in existence. Describe the key features of each and give an example.

  o **Cartilaginous fishes**:

  o **Ray-finned fishes**:

  o **Lobe-finned fishes**:

B. Movement onto Land

- While the first vertebrates lived in water, adaptation occurred that allowed groups of vertebrates to move onto land. Specifically, there are four evolutionary adaptations required for terrestrial living. Name and describe each adaptation.

  1.

  2.

  3.

  4.

V. All Terrestrial Vertebrates Are Tetrapods

- All terrestrial vertebrates have four legs and can be referred to as _____. These animals can be categorized into two major groups depending on the structure of their eggs.

  o An example of a **non-amniote** would be:

  o An example of an **amniote** would be:

A. Amphibians

- Explain why amphibians must spend at least part of their life cycle in or near water.

- Describe each of the three phases of the amphibian life cycle:

    1. Eggs:

    2. Juveniles:

    3. Adults:

B. Birds

- While it may seem like an odd grouping, birds are terrestrial vertebrates classified in the _____ group, which also includes snakes and lizards. The reason for this classification includes similarities in _____ and _____.

- Birds are unique within their group in that they have feathers and are **endotherms**, while the other animals they are grouped with are **ectotherms**. Describe the difference between the following terms.

    o Endotherm:

    o Ectotherm:

- Our understanding of the evolutionary relationship between birds and dinosaurs is changing. Describe this relationship.

- What purposes do feathers serve?

C. Mammals

- List the two features shared by all mammals.

  1.

  2.

- Describe the features of mammals that are conducive to the development of endothermy.

- Most mammals give birth to their young instead of laying eggs, which is referred to as _____. There are two subgroups of this type of mammal, which include the **marsupials** and **placentals**. Describe the adaptations of each group and provide an example animal.

  o Marsupials:

  o Placentals:

- One group of mammals does lay eggs. This group is the _____. Explain this unique adaptation and give an example animal.

D. Primates and Human Evolution

- Humans are **primates** that evolved from an arboreal lineage. Three characteristics that humans share with our arboreal primate relatives include:

  1.

  2.

  3.

- There are some key differences between the groups of primates. For example, Old World and New World Monkeys have a _____, while the apes do not have these structures. Among the ape group, the _____ and _____ live in solitary conditions or in a pair, while the _____ and _____ live in social groups.

- Figure 11-33 shows the phylogeny of the primates. Fill in the names of each group in the figure.

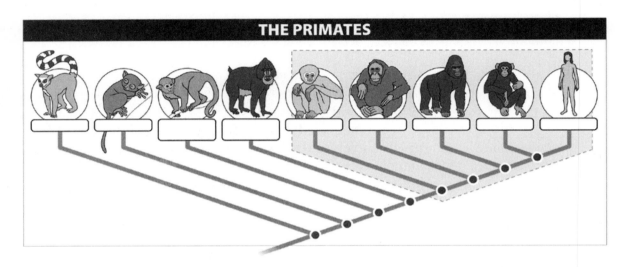

- In comparing human DNA sequences to those of existing groups of primates, we share about _____% genetic similarity with the chimpanzees. Even with this amount of genetic similarity, there are three key differences between humans and chimpanzees. Explain these differences, listing them in the order in which they evolved.

    1.

    2.

    3.

- The australopithecines evolved about _____ years ago. Describe the features that differentiated them from existing species of primates.

- From the australopithecines evolved members of the genus *Homo*. Several species in our lineage existed. Describe the key features of each.

  o *H. hablis*

  o *H. erectus*

  o *H. ergaster*

  o *H. neanderthalensis*

  o *H. floresiensis*

  o *H. sapiens*

- Figure 11-35 shows the evolutionary relationships among species of the *Homo* genus. Fill in the names of each species in their correct locations on the figure.

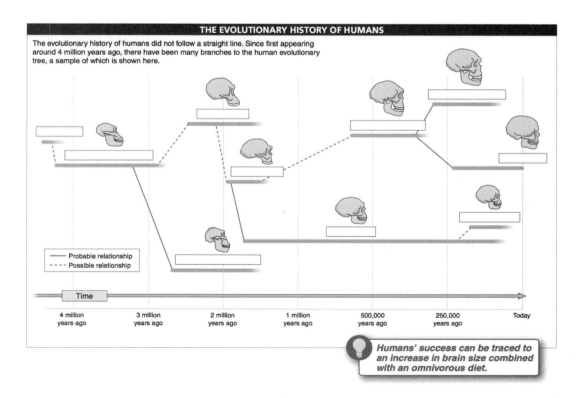

E. The Past 100,000 Years of Human Evolution

- Based on fossil and molecular evidence, the evolution of *H. sapiens* is estimated to have occurred about _____ years ago in the region of _____.

- A small group of modern *H. sapiens* likely left the area and migrated to various parts of the world. According to mitochondrial DNA analysis, three major routes of initial migration occurred. Those routes took humans to the following locations:

  1.

  2.

  3.

- As *H. sapiens* migrated through the world, they existed with other *Homo* species including *H. neanderthaensis*. The fossil record indicates that Neanderthals often sustained injuries to bones. This allows for two major deductions about them:

  1.

  2.

- Indicate which three human species co-existed with *H. sapiens*.

  o What is the most likely explanation for the extinction of the other *Homo* species?

Testing and Applying Your Understanding

Multiple Choice (For more multiple choice questions, visit www.prep-u.com.)

1. Why is the amniotic egg considered a key evolutionary innovation?
   a) It greatly increases the survival probabilities of eggs in a terrestrial environment.
   b) It extends embryonic development.
   c) It has a shell with a single membrane.
   d) It prohibits external fertilization, thereby facilitating the evolutionary innovation of internal fertilization.
   e) It enables eggs to float in an aquatic medium.

2. Which of the following series of events correctly describes the evolutionary history of the vertebrates, from earliest to latest in time?
   a) jaws : lungs : four limbs : hair
   b) four limbs : hair : lungs : jaws
   c) hair : four limbs : lungs : jaws
   d) lungs : jaws : four limbs : hair
   e) four limbs : lungs : jaws : hair

3. In cnidarians, cnidocytes are primarily used for:
   a) formation of free-living medusae.
   b) prey capture and defense.
   c) creation of water flow across the body.
   d) muscular contraction during movement.
   e) secretion of digestive enzymes.

4. Cephalization refers to:
   a) the attachment to the substrate via the head.
   b) the presence of a well-developed brain or ganglion.
   c) the placement of the head on the dorsal side of the body.
   d) the concentration of sensory organs at the end where environmental cues are first encountered.
   e) the development of the mouth on the head.

5. Both birds and mammals have four-chambered hearts. Birds, however, are more closely related to crocodiles, which have three-chambered hearts, than they are to mammals. The correct interpretation of these observations is that:
   a) crocodiles evolved from a group that originally had four-chambered hearts.
   b) the trend in heart evolution is from four-chambered to three-chambered.
   c) birds and mammals evolved four-chambered hearts independently.
   d) having a four-chambered heart would be detrimental to crocodiles.
   e) Both a) and b) are correct.

6. Which of the following traits is unique to arthropods?
   a) an exoskeleton
   b) segmentation
   c) appendages
   d) a dorsal anterior brain
   e) a ventral nerve cord

7. All groups of animals have nervous systems EXCEPT:
   a) worms.
   b) sponges.
   c) jellyfish.
   d) Neither a) nor b) have nervous systems.
   e) Neither a) nor b) nor c) have nervous systems.

8. According to the fossil record, the first humans appeared approximately:
   a) 1 million years ago.
   b) 100,000 years ago.
   c) 6 million years ago.
   d) 190 million years ago.
   e) 6,000 years ago.

9. Which is the only animal phylum to have over one million described species?
   a) chordates
   b) arthropods
   c) flatworms
   d) mollusks
   e) nematodes

10. Which came first, the chicken or the egg?
    a) The chicken, because the amniotic egg did not evolve until after the first chicken appeared.
    b) The chicken, because there had to be a chicken in order to lay an egg.
    c) The chicken, because during speciation the adult stage always precedes the juvenile stage.
    d) The egg, because the chicken is not a real species.
    e) The egg, because the amniotic egg evolved well before the first bird.

11. The two most important evolutionary innovations in vertebrates, which resulted in their eventual domination of the large animal niches, were:
    a) few predators and less competition at first.
    b) evolution of the ability to walk and fly.
    c) evolution of air-breathing lungs and the availability of terrestrial prey.
    d) evolution of air-breathing lungs and a more efficient heart.
    e) evolution of jaws and amniotic eggs.

Short Answer

1. Certain species of protozoa are often described as being "animal-like." Based on the characteristics displayed by all animals, why wouldn't a protozoan be considered a true animal?

2. Different people have different opinions on what the term "success" means in an evolutionary context. Based on what you have learned in this chapter, which animal phylum would you consider to be the most evolutionarily successful? Justify your answer by explaining how you define success.

3. Describe the relationships between body symmetry and the type of movement required by an animal.

4. An unidentified animal appears to have a notochord. Based on this information alone, to which phylum (or potential phyla) might it belong?

5. The evolution of amniotic eggs was a key adaptation that allowed animals to survive in new environments. What is an amniotic egg, and in what ways might it be advantageous to an animal?

6. Many people are surprised to learn that birds are in fact reptiles, and that they share a close relationship with the dinosaurs. Describe the evolutionary relationship between birds and dinosaurs.

7. A common misconception about human evolutionary theory is the idea that "humans evolved from monkeys." Describe the details of the human evolutionary lineage, making sure to explain our relationships to monkeys, apes, and other species.

8. Humans are unique in that we are the only remaining species in our genus, *Homo*. Given our evolutionary history, what predictions might you make for the future of *H. sapiens*?

9. The media has largely portrayed the octopus as a highly intelligent animal. Provide one piece of evidence to support that assertion and one piece of evidence that might question that assertion.

Chapter 12
PLANT AND FUNGI DIVERSIFICATION—WHERE DID ALL THE PLANTS AND FUNGI COME FROM?

Learning Objectives

- Describe how plants are different from other eukaryotic organisms
- Explain the unique features of plants that allow them to succeed on land
- Compare and contrast the features of seedless and seed plants
- Discuss the benefits of a vascular system in plants
- Describe the structure of a seed and explain ways in which seeds can be distributed
- Differentiate between the reproductive cycles of gymnosperms and angiosperms
- Discuss how flower structure relates to reproduction in angiosperms
- Compare the unique process of double fertilization in angiosperms to other groups of plants
- Explain the role of fruits in the reproductive cycle
- Discuss several ways in which plants defend themselves against predators
- Characterize the structure of a typical fungus and explain ways in which fungi may interact with other species
- Explain the reproductive cycle of a typical fungus

Chapter Outline

I. Plants Are Just One Branch of the Eukarya

- Like animals, plants are eukaryotic and multicellular. However, plants are quite different in other aspects. They produce their own food in the process of _____ and they are unable to move.

- Plants have a variety of structures specialized to meet their unique needs. Indicate the role of each of the following structures.

    o Roots:

    o Shoot:

- Not all plants contain chlorophyll. Provide an example of a plant that is not reliant on chlorophyll and explain how it acquires the nutrients it needs.

- Plants face a series of challenges based on their immobility. Explain how plants have adapted to meet each of the following challenges.

  o   Acquiring sunlight:

  o   Reproduction:

  o   Resisting predators:

- Contrast the differences between vascular and non-vascular plants.

II. The First Plants Had Neither Roots nor Seeds

- The aquatic ancestors of terrestrial plants are the _____. One type of aquatic plant that is considered to be the closest relative of land plants is the _____.

- The first terrestrial plants are thought to have appeared about _____ million years ago. Describe the features of these early plants.

- Describe the role of the cuticle in protecting land plants.

- Figure 12-4 shows the evolutionary relationships among the major groups of plants. Label each type of plant in the diagram.

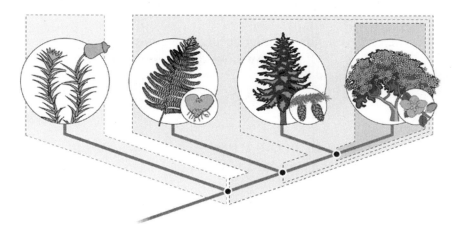

A. Non-Vascular Plants

- The non-vascular plants, also called the _____ are all low-growing. Why?

- There are three specific categories of non-vascular plants. Those include the _____, _____, and _____.

- The non-vascular plants have a method of reproduction often referred to as **alternation of generations**. Figure 12-8 depicts this life cycle. Label each structure in the spaces provided and indicate the four steps involved in this process.

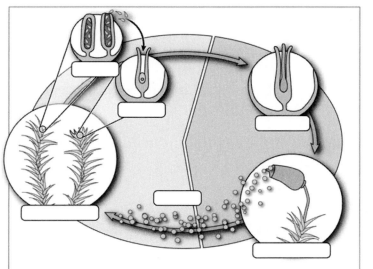

1.

2.

3.

4.

B. Vascular Seedless Plants

- The two specific categories of vascular seedless plants are the _____ and the _____. How does the anatomy of these plants relate to the environments in which they live?

- Describe how reproduction in seedless vascular plants is similar to that of the non-vascular plants.

    o   How is reproduction different between the two groups?

III. The Advent of the Seed Opened New Worlds to Plants

- The development of seeds in plants was a unique adaptation that prepared plants to survive in a more diverse set of environments. Plants that produce seeds include the _____ and _____.

- Describe how the structure of seeds is different from the spores seen in the bryophytes.

- Plants that produce seeds must produce haploid gametes. This haploid form is referred to as the _____. The female version of this form is the _____, and the male version of this form produces _____.

- Seed dispersal is a unique challenge faced by plants. List three ways that a plant might disperse its seeds.

    1.

    2.

    3.

A. Gymnosperms

- The **gymnosperms** all produce seeds but are unique in that they produce their _____ on cone-like structures.

- Describe the common characteristics of the gymnosperms:

- The four major groups of gymnosperms include:

  1.

  2.

  3.

  4.

- Cones are a unique feature to the gymnosperms. Describe how male and female cones differ from each other.

- The most common method of pollen dispersal in the pines is via _____.

- Describe the haploid stage of the life cycle in gymnosperms as it compares to that of non-vascular plants.

B. Conifers

- The **conifers** include members that are both the oldest and tallest species in existence.

- Conifers have a variety of unique features which have contributed to their evolutionary success. Describe two of those features.

    1.

    2.

IV. Flowering Plants Are the Most Diverse and Successful Plants

- Angiosperms species far outnumber those of gymnosperms. A unique adaptation of the angiosperms is that their flowers house reproductive structures seen in Figure 12-20. Label each part on the diagram.

- For each of the following structural elements of flowers seen in Figure 12-20, indicate their function and then label each on the diagram.

    o   Stamen:

    o   Anther:

    o   Filament:

    o   Carpel:

    o   Stigma:

    o   Style:

    o   Ovary:

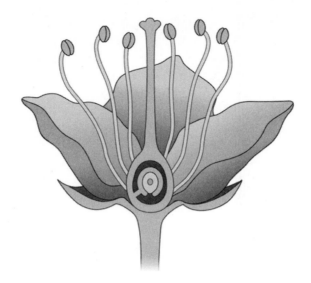

A. Pollination

- In order for pollination to occur, the male gametes must get to the female gametes. While pollination can occur through wind or water dispersal, it is more efficient to use other means. Explain how pollen is transported in these alternate means of dispersal.

    o Trickery:

    o Bribery:

- Provide several examples of animals that assist in the pollination process.

B. Double Fertilization

- Angiosperms have two separate fertilization events referred to as **double fertilization**. In this case, there are two fusions of nuclei from the pollen grain with the female nuclei.

- There are two major benefits afforded to angiosperms by double fertilization. Describe each benefit.

  1.

  2.

- Explain the measures taken to prevent self-fertilization within flowers.

- Figure 12-23 summarizes the events of double fertilization. Summarize the five major steps.

1.

2.

3.

4.

5.

- Outbreeding allows for the combination of gametes from different individuals. What benefit does outbreeding provide to plants?

V. Plants and Animals Have a Love-Hate Relationship

- The production of fruits allows flowering plants to disperse their seeds. While some fruits are dry fruits, others are fleshy fruits. Using an example of each, explain the difference between each type of fruit.

- Fruits must serve as attractive bait. Describe the three characteristics of fruits that make them appealing.

    1.

    2.

    3.

- In order to survive, plants have to find ways to encourage animals to pollinate them and discourage animals from eating them. Describe several adaptations plants have that can discourage animal predators.

- **Bioprospecting** refers to searching for compounds in plants that may have medicinal uses. Give several examples of compounds from plants that are used as pharmaceuticals.

- Explain why certain plants in low nutrient environments must supplement their meals by preying on insects.

VI. Fungi and Plants Are Partners but Not Close Relatives

- While fungi have often been thought of as more plant-like than animal-like, evidence shows that fungi are more like animals than plants. Provide evidence to support this assertion.

- The major groups of fungi that we are most familiar with include the single-celled _____ and the multicellular _____ and _____.

- A unique feature of fungal cells is that they have cell walls made of _____.

A. Life Cycles

- The structure of most fungi includes long strings of cells that can spread out over or under a surface. Those cells are called _____, and a tangled mass of these cells is referred to as a _____.

- Fungi can serve as decomposers or parasites. In either role, describe how fungi get their nutrition.

- The life cycle of a fungus can be seen in Figure 12-31. Label each blank and explain what happens in each of the six steps of the life cycle.

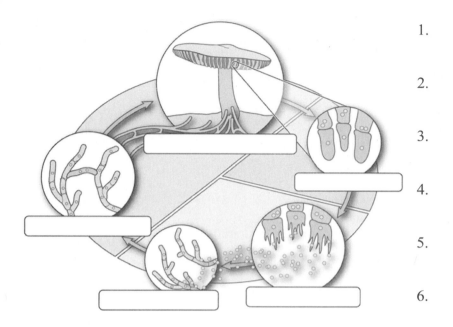

1.

2.

3.

4.

5.

6.

B. Relationships with Other Organisms

- Fungi exist in relationships with several other types of organisms. When fungi associate with cyanobacteria or algae, a _____ is formed.

- Fungi also associate with the roots of plants, forming _____.
  Provide an example of when this can be beneficial to the plant and an example of when it might be detrimental to the plant.

Testing and Applying Your Understanding

Multiple Choice (For more multiple choice questions, visit www.prep-u.com.)

1. Unlike higher plants like angiosperms and gymnosperms, all bryophytes lack:
    a) cuticles.
    b) water transport mechanisms.
    c) roots.
    d) stomata.
    e) alternation of generations.

2. Unlike the seedless plants, seed plants:
    a) undergo heterospory.
    b) have flowers and fruit.
    c) are not dependent on water for reproduction.
    d) have true roots and true leaves.
    e) have a dominant sporophyte.

3. Which of the following is the most diverse group of plants?
    a) bryophytes
    b) angiosperms
    c) nontracheophytes
    d) ferns
    e) gymnosperms

4. Angiosperms and gymnosperms differ from each other in that:
    a) angiosperms have a single fertilization process, whereas gymnosperms use double fertilization.
    b) angiosperms are heterosporous, whereas gymnosperms are homosporous.
    c) angiosperms have vessel elements instead of xylem, whereas gymnosperms have xylem.
    d) angiosperms tend to rely on animal pollinators, whereas gymnosperms tend to rely on wind pollination.
    e) angiosperms use pollen, whereas gymnosperms use cones.

5. Both mosses and ferns must have free-standing water present (e.g., water droplets) in order to fulfill their requirements for fertilization. Which of the following choices is the most likely explanation for why this would be?
    a) Both mosses and ferns have both motile male and female gametes that come together in water to fertilize.
    b) Both mosses and ferns have motile female gametes that require water to travel to the male gametes.
    c) Both mosses and ferns have motile male gametes that require water to travel to the female gametes.

d) Both mosses and ferns have gametes that require the plant to take in a lot of water in order to fulfill their photosynthetic requirements.
e) None of the above choices is a good explanation for why this occurs.

6. Carpels most likely developed from:
   a) petals.
   b) endosperm.
   c) seed cones.
   d) modified leaves.
   e) modified stems.

7. Petals:
   a) serve to attract insects to collect and disperse fruit.
   b) serve to attract animals to collect and disperse pollen.
   c) are present to protect the seeds.
   d) are present mainly to protect the stamens and carpels.
   e) serve all of the above purposes.

8. What is the one most obvious difference between the gymnosperms and angiosperms?
   a) The gymnosperms don't form flowers, whereas the angiosperms do.
   b) The gymnosperms have bark, whereas the angiosperms don't.
   c) The gymnosperms don't produce seeds, whereas the angiosperms do.
   d) The gymnosperms are tall, whereas the angiosperms are short.
   e) The gymnosperms have cones, whereas the angiosperm don't.

9. All of the traits below suggest that fungi are much more closely related to animals than they are to plants, EXCEPT:
   a) structure and position of flagella on some gametes and spores.
   b) synthesis of chitin.
   c) DNA sequences of several genes.
   d) lack of cell wall.
   e) glycogen is the food storage molecule.

10. Dispersal of fungal spores is typically accomplished by:
    a) flagella movements.
    b) ciliated movement.
    c) wind.
    d) social insects.
    e) hummingbirds.

Short Answer

1. From a survival species, is it more advantageous for a plant to spend most of its time in a haploid or diploid stage? Explain your answer.

2. Different people have different opinions on what the term "success" means in an evolutionary context. Based on what you have learned in this chapter, which plant group would you consider to be the most evolutionarily successful? Justify your answer by explaining how you define success.

3. Many people assume that carnivorous plants act like animals, eating their prey and not performing photosynthesis. Explain why carnivorous plants use insects to supplement their photosynthetic efforts.

4. Describe the process of alternation of generations in plants. What benefits come from having this sort of life cycle?

5. Which types of plants reproduce with spores? What are the advantages to these groups from using spores as their reproductive structures?

6. Suppose a plant of an unknown type seems to be producing profuse amounts of pollen. From this information, what method of pollination would you expect? Explain your answer.

7. Describe how the process of double fertilization improves the quality of seeds in the angiosperms.

8. Why is it so difficult to pinpoint the exact size of many fungi?

9. The media often presents stories of "toxic mold," which develops in buildings and makes people sick. What adaptations do these fungi have that allow them to cause human illness?

Chapter 13
EVOLUTION AND DIVERSITY AMONG THE MICROBES—BACTERIA, ARCHAEA, PROTISTS, AND VIRUSES: THE UNSEEN WORLD

Learning Objectives

- Explain how microbes are different from other groups of organisms
- Describe the typical structures of a bacterial cell and their functions
- Characterize the various methods of lateral gene transfer in bacteria
- Discuss the role of normal flora in the body and probiotic therapy
- Describe concerns related to pathogenic bacteria
- Explain the significance of antibiotic resistance in bacteria and measures humans can take to reduce the proliferation of resistant bacteria
- Characterize the features of the major groups of sexually transmitted diseases
- Compare and contrast the features of archaea with bacteria and eukaryotes
- Describe the features of protists and differentiate between the three major groups of protists
- Describe the process of viral infection and replication
- Understand the concerns with certain viruses being able to cross species boundaries to cause infection
- Explain the life cycle of the HIV virus

Chapter Outline

I. There Are Microbes in All Three Domains

- While **microbes** are found just about everywhere, we tend not to think about them much because we cannot see them. Unlike other groups of organisms that are categorized by their evolutionary relatedness, the primary criterion that distinguishes microbes from other groups of organisms is _____.

- Microbes are among some of the most evolutionarily successful organisms for the following reasons:

    1. *Microbes are genetically diverse.* Approximately _____ species of microbes are thought to exist.

2. *Microbes can live in a diverse range of settings and can acquire nutrients from diverse sources.*

- Provide several examples of environments in which microbes can live.

- What sorts of nutrients can microbes use to support their metabolic needs?

3. *Microbes are abundant.* Give a few examples of places microbes live in extremely high numbers.

- Complete the following table on the four major groups of microbes.

Microbe	Domain	Prokaryotic or eukaryotic?
Bacteria		
Protists		
Viruses		
Archaea		

II. Bacteria May Be the Most Diverse of All Organisms

- Bacteria are incredibly efficient in carrying out life's functions. They require only the following structures to function. For each of these structures, indicate its role in the bacterial cell.

  o Plasma membrane:

  o Ribosomes:

  o DNA:

  o Cytoplasm:

- In addition, there are several other optional structures that certain bacteria may possess. These include **flagella**, **pilli**, and **capsules**. Their functions include:

  o Flagella:

  o Pilli:

  o Capsules:

- Figure 13-5 illustrates typical bacterial structure. Label each of the structures in the diagram.

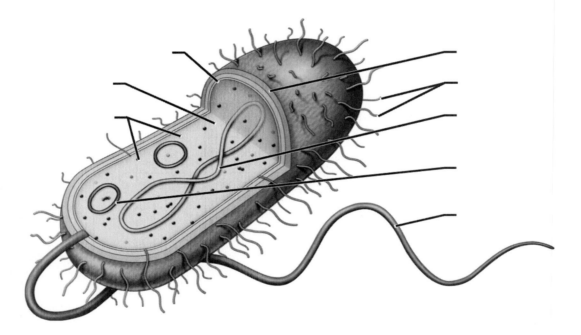

- In general, bacteria come in three shapes. Those include _____, _____, and _____.

- Major groups of bacteria can be differentiated by use of the **Gram stain**. The difference in **peptidoglycan** in the plasma membrane is responsible for the differential reaction of Gram positive and Gram negative bacteria to this stain. Explain the differences in peptidoglycan in Gram positive and Gram negative bacteria.

A. Bacterial Reproduction

- One reason bacteria are so successful is that they can divide quickly. On average and in optimal conditions, bacteria can typically reproduce every _____.

- **Binary fission**, a form of _____ reproduction, results in genetically identical daughter cells. Mutation and gene transfer are the mechanisms used to initiate genetic diversity in bacterial populations.

- Bacterial genes are located on a circular chromosome. Unlike eukaryotic chromosomes, bacterial DNA lacks introns and is organized in a way such that genes with similar functions are near each other and can be regulated as a group.

- Some bacteria have additional loops of DNA called **plasmids**. These come in several varieties. Explain the functions of each.

    o   Metabolic plasmids:

    o   Resistance plasmids:

    o   Virulence plasmids:

- In addition to passing genetic information to offspring, bacteria can perform lateral gene transfers to existing bacterial cells. For each of the following diagrams, label the type of gene transfer.

 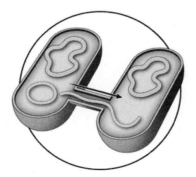 

_____   _____   _____

- Now explain how each type of gene transfer works:

  o   Conjugation:

  o   Transformation:

  o   Transduction:

B. Metabolic Diversity Among Bacteria

- Contrast the differences between each type of trophic category in bacteria:

  o   Chemoorganotrophs:

  o   Chemolithotrophs:

  o   Photoautotrophs:

- What role do cyanobacteria play in the Oxygen Revolution?

III. In Humans, Bacteria Can Have Harmful or Beneficial Health Effects

A. Beneficial Bacteria

- Bacteria have a bad reputation for causing disease. However, in reality, there are very few bacteria that cause disease. Some even help protect against disease. What other beneficial functions can bacteria serve?

- **Normal flora** are bacteria that live on and in your body. Give several examples of places you would find normal flora in or on your own body.

- While the term **probiotic therapy** sounds like a drug treatment, it involves the introduction of live bacteria to the body to provide competition for potential pathogens looking to enter. In this context, why is competition a good thing?

B. Pathogenic Bacteria

- Some bacteria do cause disease. Species that always cause disease are called _____. Other species are more opportunistic and only cause disease under certain circumstances.

    o An example of a pathogen would be:

    o An example of an opportunistic pathogen would be:

C. The Problem of Drug Resistance

- An _____drug is defined as a substance that kills bacteria. Most of these are produced by microbes themselves as a defense mechanism to out-compete other microbes.

- Bacterial resistance to **antibiotics** is conferred by the presence of specific genes present either on the bacterial chromosome or on a resistance plasmid. These genes produce proteins that have the ability to limit the effectiveness of the antibiotic in one of three ways:

    1.

    2.

    3.

- While antibiotic-resistant bacteria are inevitable, there are things we do as humans that greatly accelerate their prevalence and proliferation. Three things we could do as humans to limit the problem of antibiotic-resistant bacteria include:

    1.

    2.

    3.

D. Sexually Transmitted Diseases

- List the types of microbes that may cause an STD:

- Explain the two reasons that STDs are nearly impossible to eradicate from a population:

    1.

    2.

- Make sure to review the features of the most common STDs listed in Figure 13-5.

IV. Archaea Exploit Some of the Most Extreme Habitats

- For quite some time, **archaea** were thought to be bacteria because, superficially, they do have some features in common with each other. What we now know is that although the two groups look quite similar externally, they are quite different from each other internally.

    o   Indicate two characteristics that distinguish archaea from bacteria.

    o   Indicate two characteristics that distinguish archaea from eukaryotes.

- Archaea (and some bacteria) are known as **extremophiles**. Provide several examples of the types of "extreme" habitats these organisms may occupy.

- Because of extreme conditions under which some archaea can survive, they may have potential industrial and environmental application. Provide an example:

V. Most Protists Are Single-Celled Eukaryotes

- Protists are approximately _____ times larger than bacteria and archaea.

- Describe the internal characteristics that differentiate protists from other microbes.

- Some organelles in the protists have double membranes. Explain how these organelles evolved to have double membranes.

    o   Nucleus:

    o   Mitochondria:

A. Types of Protists

- Protists are a diverse group of eukaryotic microbes. They are categorized as: animal-like, plant-like, or fungus-like. Fill in the following table to compare the differences between the major groups of protists.

	Animal-like	Plant-like	Fungus-like
Example organism			
Typical habitat			
Feeding mechanism			
Unique features			

B. Pathogenic Protists

- Pathogenic protists are known as _____, which spend at least part of their life cycle living on/in a _____.

- *Plasmodium* is an example of a pathogenic protist that infects red blood cells during a portion of its life cycle and causes the disease _____. Provide a basic description of the *Plasmodium* life cycle.

- Some people are resistant to infection by *Plasmodium*. Explain how this can occur.

VI. Viruses Are at the Border between Living and Non-Living

- Viruses consist of genetic material, either _____ or _____, which is wrapped in a protective _____.

- In order for a virus to successfully infect a host cell, the virus must interact with a _____ on the surface of the host cell. Without this interaction, the virus will not infect the cell.

- The basic steps of viral replication are shown in Figure 13-25. Fill in the events of the main steps labeled 1–5 on the diagram.

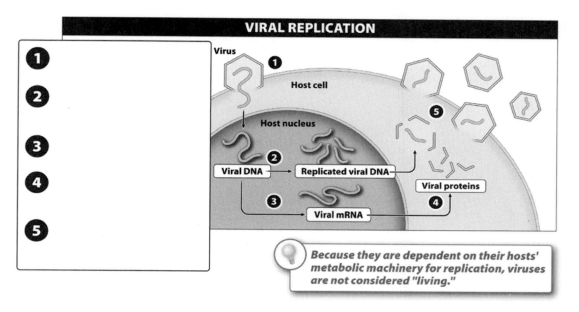

VIRAL REPLICATION

1 ___
2 ___
3 ___
4 ___
5 ___

Virus
Host cell
Host nucleus
Viral DNA → Replicated viral DNA
Viral mRNA
Viral proteins

Because they are dependent on their hosts' metabolic machinery for replication, viruses are not considered "living."

A. Viral Infections

- Some viruses are merely nuisances while others can cause fatal infections. When a virus causes a worldwide epidemic, this is referred to as a _____. An example of this would be:

- While certain viruses can be prevented by vaccination, others cannot. In particular, viruses that use _____ as their genetic material tend to change frequently making vaccination efforts futile. Why is this the case?

- Most viruses are specific to a single type of host cell that they can infect. However, there are viruses that can infect multiple types of cells and even multiple species. Using Influenza A as an example, explain how this virus can jump from species to species.

B. The HIV Virus

- HIV belongs to a unique category of viruses called the **retroviruses** which have the ability to code for the enzyme _____, which converts viral RNA to DNA.

- One reason that HIV is so difficult to treat is because it mutates at an incredible rate. What feature of the virus is responsible for its frequent mutation?

- Explain what must happen with the HIV virus for it to trigger the development of AIDS in its host.

- The life cycle of the HIV virus can be seen in Figure 13-29. Label the appropriate events of each stage in the diagram.

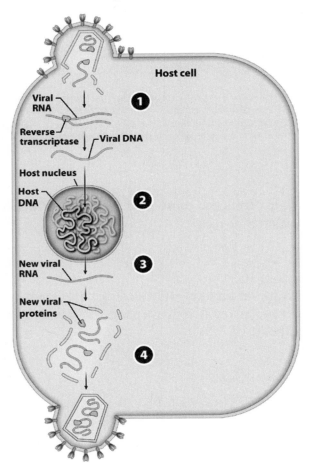

Testing and Applying Your Understanding

Multiple Choice (For more multiple choice questions, visit www.prep-u.com.)

1. Protists are alike in that they all are:
    a) multicellular.
    b) eukaryotic.
    c) nonparasitic.
    d) marine.
    e) photosynthetic.

2. To what does the phrase "horizontal gene transfer" refer?
    a) the only method of genetic recombination in bacteria
    b) the mechanism of endosymbiosis
    c) the transfer of genetic material information from one individual to another that is not
        its offspring
    d) recombination between individuals with linear (rather than circular) chromosomes
    e) All of the above.

3. The basic structure of a virus consists of:
    a) genetic information in the form of DNA, a protein coat, and polymerase to replicate
        its DNA.
    b) genetic information in the form of DNA.
    c) genetic information in the form of DNA or RNA enclosed within a protein coat, along
        with a few enzymes.
    d) genetic information in the form of DNA or RNA enclosed within a protein coat.
    e) genetic information in the form of RNA, a protein coat, and reverse transcriptase in
        order to turn its RNA to DNA.

4. How are viruses able to recognize a host cell and enter it?
    a) Viruses enter through the pores already present on a host cell.
    b) Viruses trick the cell into engulfing it.
    c) Viruses attach to host cell surface molecules that were created for some other
        function of the cell.
    d) Viruses have spikes outside of their protein coats or envelopes that puncture the
        outside membrane of the host cell and the virus enters through the hole.
    e) Viruses adhere to proteins on the host cell's surface that were created specifically for
        viral entry.

5. All prokaryotes reproduce by:
    a) budding out from host cells.
    b) sexual reproduction.
    c) endospore production.
    d) binary fission.
    e) All of the above are possible ways that prokaryotes reproduce.

6. Viruses are often used in laboratories as vectors for gene therapy. Which of the following properties makes viruses so useful in these types of experiments?
   a) Viral enzymes inject the genetic material into the host cell genetic material at exactly the right place in the DNA.
   b) Viruses are powerful replicating machines that will produce many copies of any genetic material put into their capsids.
   c) Viruses do not have their own genetic material and are thus empty shells waiting to be filled.
   d) Viruses are very adept at entering host cells and will inject any genetic material inside their capsid into a host cell.
   e) None of the above.

7. How do plasmids containing genes for antibiotic resistance get exchanged between different bacteria cells in a culture?
   a) through conjugation
   b) through artificial exchange
   c) through cloning
   d) through conduction
   e) through transduction

8. Which of the following is the correct step-by-step description of the most basic viral replication cycle?
   a) Virus binds receptor; nucleic acid is released; enzymes transcribe RNA into DNA; DNA enters host DNA; host machinery is used to replicate viral nucleic acids and proteins; new viruses are assembled and leave host cell.
   b) Virus binds receptor; virus enters host cell; nucleic acid is released; host machinery is used to replicate viral nucleic acids and proteins; new viruses are assembled and leave host cell.
   c) Virus binds receptor; virus enters host cell; host machinery is used to replicate viral nucleic acids and proteins; new viruses are assembled and leave host cells.
   d) Virus binds receptor; nucleic acid is released; viral DNA enters host DNA; host machinery is used to replicate viral nucleic acids and proteins; virus progeny buds from host cell.
   e) Virus binds receptor; nucleic acid is released; viral RNA enters host DNA; host machinery is used to replicate viral nucleic acids and proteins; virus progeny buds from host cell.

9. Which of the following statements about prokaryotes is INCORRECT?
   a) Prokaryotes have circular pieces of DNA within their nuclei.
   b) Prokaryotes appeared on earth prior to eukaryotes.
   c) Prokaryotes contain ribosomes.
   d) Some prokaryotes can conduct photosynthesis.
   e) Prokaryotes contain cytoplasm.

Short Answer

1. In this chapter you learned that your body is carrying around about 100 trillion cells but only 10 trillion of those cells belong to you. What are the remaining 90 trillion cells and how can your body support carrying around so many additional cells?

2. One category of antibiotics is referred to as broad spectrum. This means that these drugs are capable of killing or halting the replication of many bacterial species. When people take broad spectrum antibiotics, they can have side-effects such as digestive upset and intestinal distress. Speculate on why this may be the case.

3. You learned about photosynthesis and its oxygen producing abilities in Chapter 4. In many areas there are deciduous forms of vegetation that lose their leaves and halt photosynthesis during the winter. This might appear to cause a drastic reduction in the amount of oxygen available. However, there is another source of oxygen available that you learned about in this chapter. What is that source of oxygen?

4. We probably all know someone that insists on seeing a physician and procuring antibiotics at each and every sign of the slightest sniffle. You may even know someone who doesn't finish all of their prescribed antibiotics, but instead saves them for a later time to self-medicate. Why is this sort of behavior problematic? To what sorts of problems does it contribute?

5. For many years, archaea were confused with bacteria. We now know that while archaea do have some features in common with bacteria, they are in fact quite different from bacteria. Explain two similarities and two differences between them.

   • Similarities:

   • Differences:

6. You have a friend that insists on constantly using antibacterial products—hand sanitizer, body wash, shampoo, etc. Even with this behavior she seems to get sick frequently. Explain to your friend why she could possibly be causing herself more harm than good by overusing these products.

7. A patient suffering from an infection goes to the physician. The physician recommends culturing to determine the cause of the infection. After the culturing has been performed, the infection is found to be caused by unicellular microbes with a nucleus. Based on this information, what type of microbe is causing the infection? Justify your answer.

8. Viruses are a unique biological entity in that they can carry on the functions of life, but only when they enter into a relationship with a host cell. Using the characteristics of life you learned about in Chapter 1, justify whether you think a virus is living or not.

9. For years we have seen news stories that talk about the potential development of an HIV vaccine. In reality, making a vaccine for HIV has proven extraordinarily difficult. Given what you know about the HIV virus, explain why a vaccination is so difficult.

Chapter 14
POPULATION ECOLOGY: PLANET AT CAPACITY—
PATTERNS OF POPULATION GROWTH

Learning Objectives

- Understand the fundamentals of ecology
- Identify the key perspective needed in population ecology
- Identify and explain the important factors that impact population growth
- Outline types of population growth
- Understand the variations of life histories
- Define survivorship curves and the relationship between growth, reproduction, and survival
- Understand aging and why some organisms age faster than others
- Outline experiments that demonstrate how longevity can be extended
- Understand the information provided in an age pyramid
- Define an ecological footprint
- Describe the current state of human population growth

Chapter Outline

I. **Population Ecology Is the Study of How Populations Interact with Their Environments**

A. **What is Ecology?**

- **Ecology** can be defined as the study of:

- This involves different or varying levels from individuals to ecosystems. Define the following levels and give an example.

  o Individuals

  o Populations

      ○   Communities

      ○   Ecosystem

- How populations interact with their environment—including the influence of other species, the environment, and their growth—can be referred to as a subfield of ecology called _____.

      ○   The important processes in population ecology include adaptations, birth rates, death rates, immigration, and emigration. Can any of these processes be studied at the individual level? Explain.

B. Population Growth

- How does a population remain stable (neither growing nor shrinking)? Explain.

- Define the **growth rate** of a population.

- If *r* stands for growth rate, give the equation necessary to calculate the growth of a population in a year.

- Define **exponential growth**.

      ○   Can this type of growth occur for long periods of time? Explain.

C. Factors that Affect a Population's Growth

- A population's growth is impacted by the number of individual organisms in a specific area, or the **population density**.

- List four **density-dependent factors** that can impact growth.

  1.

  2.

  3.

  4.

     o Can you come up with another factor of your own?

- **K** represents _____ _____ of the environment.

  o List two related consequences that could occur if a population size approached its "**K**."

     1.

     2.

- Explain what occurs during **logistic growth**. How is this different than exponential growth?

- Explain how **density-independent factors** impact a population, and give an example of one such factor.

D. Population Cycles

- The _____ growth pattern is the best model to outline the expected growth of populations; however, it is important to understand how certain growth patterns can deviate from the expected.

- There are some important examples that illustrate instances when populations do not follow the expected patterns of growth. Explain the pattern of growth for each of the following populations and explain how this cycle occurred.
    - Desert locust population in northwest Africa

    - Lynx and snowshoe hare populations in Canada

- What population growth cycle is a bold urban legend?

E. Concept of Maximum Sustainable Yield

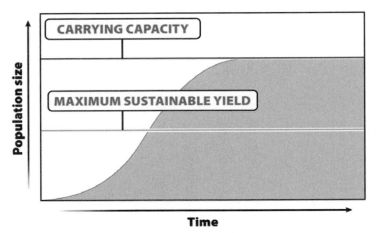

- An important concept when studying the management of natural resources is the **maximum sustainable yield**. Give a brief definition of this term.

    - While this concept is intended as a positive long-term management solution, it relies on knowing the population's _____.
    - Why is it difficult to estimate or calculate this value?

II. A Life History Is Like an Executive Summary of a Species

A. Life Histories Are Shaped by Natural Selection

- The executive summary of an organism, or its **life history,** includes important statistics such as:

  - ○

  - ○

  - ○

  - ○

- We can observe variations in the life histories of different organisms. One aspect that can vary is the organism's **reproductive investment**. Describe, in general terms, an organism's reproductive investment.

- Explain the two factors to consider when evaluating the costs and benefits of one reproductive strategy over another.

  1.

  2.

    - ○ As an example, explain how humans and rodents differ in life histories.

B. Populations Can Be Described Quantitatively in Life Tables and Survivorship Curves

- Although life tables are used when an individual is purchasing life insurance, biologists can use the information in an organism's life table to create an important graph called a **survivorship curve**. The information from a life table includes:

  - ○ Longevity, or how long the organism is expected to live

  - ○

  - ○

- What, specifically, is plotted on a **survivorship curve**?

- The information plotted can be broken down into three specific areas or types. Explain what each type means.

  1. Type I

  2. Type II

  3. Type III

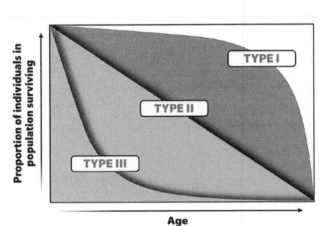

C. There Are Tradeoffs between Reproduction and Longevity

- An organism's fitness can be impacted by areas of its life history such as *growth*, *reproduction*, and *survival*.

- There are three major tradeoffs in the investments of an organism's life history. Indicate and explain each of these tradeoffs.

  1.

  2.

  3.

III. Ecology Influences the Evolution of Aging in a Population

A. Things Fall Apart: What is aging and why does it occur?

- If you asked different individuals to define aging, you would hear different definitions. Therefore, we need to define **aging** based on a population. Provide this information below.

- While the average lifespan in the United States is _____ years, the oldest human ever documented lived to be at least _____ years old.

- Explain the relationship between natural selection and why we age.

- Briefly define reproductive output.

- If a person is carrying a mutation that will cause disease or death at a very young age, is it possible for him or her to pass the mutation on to future offspring? Explain.

- If a person is carrying a mutation that will cause disease or death later in life, is it possible for him or her to pass the mutation on to future offspring? Explain.

- Even though there are many causes of death, the accumulation of "bad" alleles is responsible for the physical breakdown of an individual.

B. What Determines the Longevity of Different Species?

- A key factor in understanding why some organisms age more rapidly than others is the age at which the organism reproduces. Explain why.

- The world can often be a scary place, and some organisms live in a much riskier environment than others. Describe **hazard factors** and give an example.

- Explain the connection between an organism's hazard factor and reproduction.

C. Can We Slow Down the Process of Aging?

- Outline the experimental steps researchers took that resulted in doubling the lifespan of fruit flies.

- Explain the impact of delaying the fruit fly egg collection in the process of starting a new generation of flies.

IV. The Human Population is Growing Rapidly

A. Age Pyramids Reveal Much about a Population

- How does a "baby boom" impact society?

- Describe an "age pyramid" and its use.

- What is the shape of the age pyramid in the United States?

  o What is the concern over the shape of this pyramid?

B. From Third World to First World: As less-developed countries become more developed, a demographic transition often occurs.

- Population growth is important to measure and track as it impacts public health, food production, social services, and other facets of our society.

- Define **demographic transition**

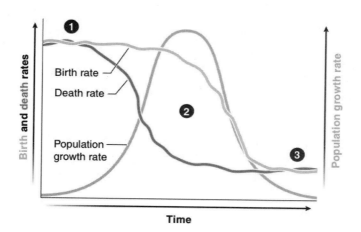

- Describe all the possible combinations of birth and death rates when the population growth rate is slow. Describe the possible combinations when the population growth rate is fast.

- Give an example of a country or area where the demographic transition has been completed.

- Give an example of an area that is currently going through this transition.

C. Human Population Growth: How High Can It Go?

- Can the carrying capacity of humans be accurately established? Explain why or why not.

- Our population is estimated at over _____ billion people. The United Nations estimates that our carrying capacity is between _____ and _____ billion.

- List three ways that humans, with the aid of technology, can increase our carrying capacity.

    1.

2.

3.

- What information does an **ecological footprint** provide?

    o   Is the ecological footprint the same from country to country? Explain.

- Provide some of your own thoughts of what impact humans may feel if the world population reaches our carrying capacity on earth.

Testing and Applying Your Understanding

Multiple Choice (For more multiple choice questions, visit www.prep-u.com.)

1. The demographic transition:
   a) is not influenced by the quality of health care available to a population.
   b) occurs in predator-prey pairs, such as the lynx and hare, whose population sizes cycle regularly.
   c) includes a decrease in the birth rate followed by a decrease in the death rate as a population becomes industrialized.
   d) is characterized by an increase in a population's growth rate.
   e) includes a decrease in the death rate followed by a decrease in the birth rate as a population becomes industrialized.

2. How does exponential growth differ from logistic growth?
   a) With logistic growth generally comes infinite expansion.
   b) Exponential growth models include consideration of a population's carrying capacity.
   c) Long-term exponential growth is more commonly observed than long-term logistic growth in nature.
   d) The logistic model of growth incorporates environmental limitations on population size.
   e) Logistic growth models take the population's age-structure into account.

3. A primary difference in the age pyramids of industrialized versus third-world countries is:
   a) mean longevity is significantly greater among third-world countries.
   b) third-world countries show a characteristic "bulge," which indicates a baby boom.
   c) third-world countries have significantly more individuals than industrialized countries.
   d) third-world countries have much larger proportions of their population in the youngest age group.
   e) in third-world countries females live significantly longer than men, whereas in industrial countries the reverse is true.

4. A Type-III survivorship curve would be expected in a species in which:
   a) biparental care is necessary.
   b) mortality occurs at a constant rate over the lifespan.
   c) parental care is extensive.
   d) mortality rate is quite low for the young.
   e) a large number of offspring are produced but parental care is minimal.

5. A population is:
   a) a group of individuals of the same species that live in the same general location and have the potential to interbreed.
   b) a group of individuals of related species that live in the same general location and have the potential to interbreed.

    c) a group of individuals of the same species that live in the same general location and have the same genotypes.
    d) a group of individuals of the same species that have the potential to interbreed.
    e) a group of species that share the same habitat.

6.   Which of the following is NOT an example of a density-dependent limiting factor that will influence carrying capacity?
    a) food supply
    b) predation
    c) disease
    d) flooding
    e) territory availability

7.   In years when beech trees produce a large crop of nuts, growth rings are narrow. This is best explained by:
    a) the tradeoff between the number and size of offspring an organism can produce.
    b) their decreased reproductive value.
    c) the Type-I life history of beech trees.
    d) density-dependent regulation affects.
    e) the tradeoff that exists between growth and reproduction.

8.   In a population, as N approaches K, the logistic growth equation predicts that:
    a) the carrying capacity of the environment will increase.
    b) the growth rate will approach zero.
    c) the carrying capacity of the environment will decrease.
    d) the population size will decrease.
    e) the population size will increase exponentially.

9.   The number of individuals that can be supported in a given habitat is the:
    a) density-independent effect.
    b) innate capacity for increase.
    c) density-dependent effect.
    d) biotic potential.
    e) carrying capacity.

10. Approximately what is the current size of the human population?
    a) 10.1 billion
    b) 5.8 billion
    c) 6.7 billion
    d) 85.6 billion
    e) 2.2 billion

Short Answer

1. Compare and contrast how density-dependent factors and density-independent factors impact populations.

2. Explain how technology can increase carrying capacity in some environments as well as potentially decrease the carrying capacity in others.

3. Many populations cycle between periods of growth. What is the biggest difference between the growth cycles of the desert locust and the lynx/snowshoe hare populations?

4. What are some of the important characteristics an organism must possess if they fall into the Type I group on the survivorship curve?

5. As humans, we often look back toward our parents, grandparents, and even great-grandparents to see how long we might live. Why do we age, and what is the familial, or genetic, connection to aging?

6. Indicate and explain the potential differences in hazard factors between a population of box turtles that can live up to 100 years of age and a population of bullfrogs that live an average of eight years.

7. The ecological footprints of individuals living in different countries will vary greatly depending on the country. As the text indicated, the ecological footprint of people living in India is much smaller than that of people living in Australia or Norway. How does the ecological footprint impact the earth's carrying capacity?

8. In your opinion, what should be done about the exploding population—increase the carrying capacity via technology, reduce the ecological footprint of wealthier nations, impose a mandatory population control policy, or something else? Back up your argument with information from the text and outside resources.

Chapter 15
ECOSYSTEMS AND COMMUNITIES—ORGANISMS AND THEIR ENVIRONMENTS

Learning Objectives

- Define what makes up an ecosystem
- Compare and contrast terrestrial biomes and aquatic biomes
- Outline how the earth's shape, topography, and ocean currents all impact weather patterns
- Describe how energy flows through an ecosystem
- Define each of the three important chemical cycles in an ecosystem
- Understand how species coevolve
- Define the impact of an organism's niche
- Understand the role of competition within a niche
- Outline the characteristic adaptations of predators and prey
- Explain alternative relationships in an ecosystem such as parasitism, mutualism, and commensalism
- Identify keystone species and the specific manners of change within a community over time

Chapter Outline

I. Ecosystems Have Living and Non-Living Components

A. What are ecosystems?

- Ecosystems can be of various sizes involving all different living organisms.

- What are the two specific features necessary to construct an ecosystem? Be sure to include **community** and **habitat** in your explanation.

  1.

  2.

     o What is the difference between **biotic** and **abiotic** environments?

- o   Give some examples of the *chemical resources* found in an ecosystem.

- o   Give a few examples of the *physical conditions* that impact an ecosystem.

- What is learned from biologists measuring and analyzing activities within an ecosystem?

B. A Variety of Biomes Occur Around the World, Each Determined by Temperature and Rainfall

- What are **biomes**?

- **Desert**, **savanna**, and **tropical forest** are all examples of _____ biomes.

- When defining these types of biomes biologists aren't focusing on the location, but rather the weather. What two specific aspects of the weather are evaluated when determining the type of biome?

   1.

   2.

- These aspects impact the **primary productivity** or the organic matter produced in the biome. What is the term for the organisms responsible for the **primary productivity** of a biome? Give an example.

- **Estuaries**, coral reefs, and rivers can all be classified as _____ biomes.

   - o   This classification is based on, among other things, the following features:

      -
      -
      -

II. Physical Forces Interacting Create Weather

A. Global Air Circulation Patterns Create Deserts and Rainforests

- The earth's shape influences a distribution of varying temperatures. Varying temperatures in turn influence rainfall.

- The equator region of the globe receives direct solar energy. However, the same solar energy hits the North Pole and South Pole. Explain why, specifically, it is so much cooler in these areas as opposed to the equator.

- In evaluating the ability of air to hold moisture, what body of air would hold more moisture: a warm body of air or a cold body of air?

   o Explain the process of rain formation.

- What is unique about the areas of the globe that lie at about 30 degrees latitude (either north or south)?

B. Local Topography Influences the Weather

- **Topography** can be defined as:

- There are two kinds of local topography: natural features and human engineered features.

- Natural features include:
1. Altitudes
   o Why is it when altitude increases, temperature conversely decreases?

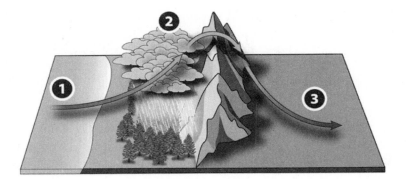

2. Deserts as a result of rain shadows

　　　o　Using the graphic above, describe the rain shadow effect.

- Human engineered features include:

1. Asphalt and cement. Urban areas with large amounts of these features see an increase in heat. Explain why.

2. Building height. Urban areas with many tall buildings experience higher winds. Explain why.

C. Ocean Currents Affect the Weather

- A combination of forces allows the ocean current to be in continuous circulation. List at least three of these forces.

- An important characteristic of water, part of the reason why it is essential to life on earth, is its ability to absorb and hold heat. How does this compare to air and how would it affect coastal temperatures?

- List the parts of the globe where the Gulf Stream current circulates.

- A change to the surface temperature in the region of the central Pacific Ocean can actually cause extreme changes to weather patterns in other areas. This change or disruption is referred to as _____.

   o Describe how a year with this change is different than a typical year.

III. Energy and Chemicals Flow within Ecosystems

A. Energy Flows from Producers to Consumers

- The flow or pathway energy takes in an ecosystem reflects important characteristics of that particular ecosystem. Regardless of the type of ecosystem, the energy to support all life comes from the _____.

- Solar energy undergoes many transformations as it passes through an ecosystem. The various levels are referred to as _____ levels.

- Define each level of energy flow and give an example.
   o **Primary producers**

   o **Primary consumers**

      ▪ What is an **herbivore**?

   o **Secondary consumers**

      ▪ What is a **carnivore?**

   o **Tertiary consumers**

- Why are **food chains** often more appropriately described as **food webs**?

- What is the role of a detritivore in the food chain? Give an example of this organism.

B. Energy Pyramids Reveal the Inefficiency of Food Chains

- Living organisms, including carnivores and herbivores, can only convert ____% of what they consume into their own mass. The remaining ____% is diverted to what?

- In understanding the flow of energy in an ecosystem, it is important to define the **biomass** of the given area. What does **biomass** refer to?

- Describe what an **energy pyramid** depicts.

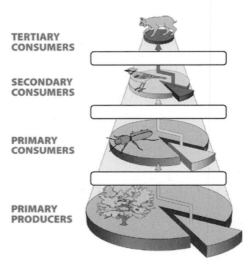

TERTIARY
CONSUMERS

SECONDARY
CONSUMERS

PRIMARY
CONSUMERS

PRIMARY
PRODUCERS

- The relative shape of the pyramid can also vary depending on the ecosystem. Do the primary producers or the top consumers more greatly impact the shape of the pyramid? Why?

C. Essential Chemicals Cycle through Ecosystems

- What is required to sustain life in an ecosystem? When you boil it down, there are two required items: _____ and _____.
    - These include three important molecules that cycle through the ecosystem.
        1.
        2.
        3.

- Regardless of the molecule's specific cycle, briefly describe the general pathway a molecule takes as it moves through the ecosystem.

- In order to define each of the three important chemical cycles, indicate the following for each cycle: give an example of how the chemical is utilized, why the chemical is necessary, and indicate the specific reservoirs in which it is found.

  1.

  2.

  3.

- Humans can disrupt some of these important cycles. **Eutrophication** occurs when:

IV. Species Interactions Influence the Structure of Communities

A. Interacting Species Evolve Together

- The reproductive success of an organism involves many selective pressures. Does natural selection rely on biotic forces, abiotic forces, or both? Explain.

- This can result in organisms evolving together or having **coevolved**.

B. Each Species' Role in a Community Is Defined by Its Niche

- A particular species with in a *community* often has its own **niche**. However, there are some environments where organisms will seek out similar resources and have competition within their own niches.

- List the characteristics that define an organism's niche.

    1.

    2.

    3.

    4.

- External pressures can influence the environmental conditions and range within which an organism can inhabit its niche. Explain the difference between a **realized niche** and a **fundamental niche**.

C. Competition Can Be Hard to See, but It Still Influences Community Structure

- When the niches of different species overlap, the competition can lead to **competitive exclusion** or **resource partitioning**.

- Indicate what occurs in each environment and the expected outcome.

    1. **Competitive Exclusion**

    2. **Resource Partitioning**

        o   What is **character displacement**?

- In summary, competition can influence a community by forcing a species to leave the area or become extinct in that area, or even result in character displacement.

D. Predation Produces Adaptation in Both Predators and Their Prey

- Define **predation**.

  o Give an example of predation that doesn't involve killing the prey.

- Predators have the ability to influence, as a selective force, the adaptations of their prey. Prey have both *behavioral* and *physical* adaptations that allow them to defend themselves against their predators.

  o List and define the *physical defenses*.

      1.

      2.

      3.

      4.

  o List and define the *behavioral defenses*.

      1.

      2.

- On the other hand, some organisms have evolved features that allow them to be better or more efficient predators. Give an example.

E. Parasitism Is a Form of Predation

- Normally if we heard the word "parasite" we would think of some type of "bug." Parasites are organisms that are much more complex than just "bugs"; they vary in their shape and size and have certain features that distinguish them from other organisms.

- **Parasitism** is a relationship between two organisms in which one particular organism (a **host**) is _____, while the **parasite** _____ (from the relationship).

    o What does symbiotic mean?

- Two different categories of parasites include **ectoparasites** and **endoparasites**. Describe where each type of parasite maintains its relationship and give an example.

    1. Ectoparasites

    2. Endoparasites

- It is essential for parasites to have access to new hosts after the relationship with its original host has run its course. The text provides several specific examples of how this is accomplished. In your own words, briefly describe one case.

F. Not All Species Interactions Are Negative

- Finally, there are specific relationships in which harm does not befall the organisms engaged in the particular interaction.

    o Define **mutualism** and give an example.

    o Define **commensalism** and give an example.

V. Communities Can Change or Remain Stable Over Time

A. Many Communities Change Over Time

- A more appropriate term for a certain type of change in a community is **succession**. Define **succession**.

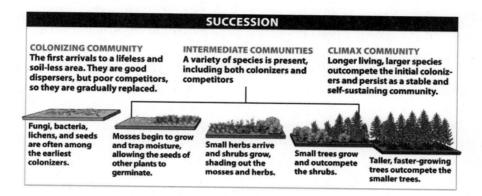

- **Primary succession** does not take place over a short period of time. It occurs after a disturbance so significant there is no life and no soil left to start over. Make a list, in order of progression, of the organisms you would see appearing and growing in an area that previously contained no life.

- Describe how **secondary succession** differs from primary succession.

B. Some Species Are More Important than Others within a Community

- In efforts to preserve the integrity and biodiversity of communities it is valuable to identify the very influential species referred to as the **keystone species**.

- Using one of the examples from the text, explain what happens when a keystone species is incorporated into the community. Then explain what happens when that species is removed from the community.

Testing and Applying Your Understanding

Multiple Choice (For more multiple choice questions, visit www.prep-u.com.)

1. Every place on earth receives the same number of hours of sunlight each year—an average of 12 hours per day. Why do different areas of the earth receive different amounts of solar radiation?
   a) The height of a place above sea level is an important determinant of the amount of solar energy that place receives.
   b) The angle of the sunlight is an important determinant of the amount of solar energy received.
   c) The density of the vegetation affects how much solar radiation a place will receive—the more grassland, the less solar radiation.
   d) The density of the vegetation affects how much solar radiation a place will receive—the more tall trees, the less solar radiation.
   e) The winds of the upper stratosphere can blow solar radiation away from an area before it reaches the land.

2. Which of the following is NOT one of the major biogeochemical cycles?
   a) hydrogen cycle
   b) carbon cycle
   c) hydrological cycle
   d) nitrogen cycle
   e) phosphorus cycle

3. The 10% rule of energy conversion efficiency:
   a) explains why herbivore biomass must exceed that of carnivores.
   b) suggests that 90% of what an organisms eats is used in cellular respiration or is lost as feces.
   c) limits the length of food chains.
   d) explains why big, fierce animals are so rare.
   e) All of the above are correct.

4. Which of the following statements about an organism's niche is NOT true?
   a) It is not always fully exploited.
   b) It includes the type and amount of food it consumes.
   c) It may be occupied by two species, as long as they are not competitors.
   d) It reflects the ways in which the organism utilizes the resources of its environment.
   e) It encompasses the space it requires.

5. Coevolution:
   a) can produce an insect with a tongue as long as its body.
   b) is responsible for all of the beautiful flowers in the world.
   c) is responsible for all nectar production by plants.
   d) reveals that both biotic and abiotic resources can serve as selective forces.
   e) All of the above are correct.

6. Unlike your cat, termites are able to use wood as their main source of energy. The best explanation for this ability is:
a) their digestive systems produce the enzyme cellulase, necessary to digest the cellulose.
b) they use the non-cellulose portion of the wood, thus having to eat continuously.
c) their digestive systems contain symbiotic bacteria, which produce a cellulose-digesting enzyme.
d) they are slow moving and do not need very much energy.
e) they have very large chitonous teeth capable of masticating the wood into cud, which is digestible.

7. A keystone predator increases species diversity when it preys upon _____ and decreases diversity when it preys upon _____.
a) arnivores; herbivores
b) competitively superior species; competitively inferior species
c) primary producers; consumers
d) herbivores; carnivores
e) competitively inferior species; competitively superior species

8. The relationship between an alga and a fungus that constitutes a lichen BEST can be characterized as:
a) a commensalism and a symbiosis.
b) a mutualism but not a symbiosis.
c) competition.
d) a mutualism and a symbiosis.
e) a symbiosis but not a mutualism.

9. Secondary consumers are:
a) detritivores.
b) herbivores that eat detritovores.
c) carnivores that eat herbivores.
d) autotrophs.
e) carnivores that eat other carnivores.

10. The total dry weight of all organisms in an ecosystem is called its:
a) flow-through energy.
b) net productivity.
c) biodiversity.
d) productivity.
e) biomass.

Short Answer

1. In your own words, define ecosystem and give a specific example of an ecosystem you can observe where you live.

2. What are the differences in determining a terrestrial biome versus an aquatic biome? Can you find multiple biomes in one geographic location? Explain.

3. The average water temperature off the coast of Miami, Florida in January is 71°F. The average water temperature in January off the coast of San Diego, California is 58°F. Explain why the water is so much warmer off the coast of Florida (and therefore more appealing for those looking to vacation in January) versus off the coast of California.

4. Regarding the degree of efficiency of energy transfer in a food chain, where does the remaining 90%, which is not biomass, "go"? Explain.

5. Explain in what ecosystem a small-based pyramid would be found.

6. Even though some parasites can hardly be seen by the naked eye, explain why they can be considered predators.

7. In your own words, distinguish between commensalism and mutualism.

8. In regards to keystone species:
   a. Why is the identification of a keystone species so important?
   b. After doing a little research, explain the role of a sea otter in its ecosystem. Here are some questions to get you started: What is its relation to predators and to prey in its environment? Do we seen an impact if the sea otter has been removed from its environment? Explain.

Chapter 16
CONSERVATION AND BIODIVERSITY—HUMAN INFLUENCES ON THE ENVIRONMENT

Learning Objectives

- Understand the unique values biodiversity has to humans
- Identify how biodiversity is distributed around the globe
- Understand the island biogeography theory
- Distinguish between types of extinctions
- Identify causes of extinctions
- Distinguish between ecosystem disturbances that are reversible and those that are irreversible
- Identify and understand the key events that lead to significant ecosystem disturbance and change
- Outline current strategies for preservation

Chapter Outline

I. Measuring and Defining Biodiversity Is Complex

A. Biodiversity Benefits Humans in Many Ways

- There are many organisms that are of great medical value to humans; other organisms of value have the ability to decompose unwanted human waste. This type of value, or view of **biodiversity**, can be considered _____ (a narrow view).

- Humans value biodiversity in other ways as well. Describe the three types of value that differ from the viewpoint above.

B. Biodiversity Is Not Easily Defined

- It is often easiest to define biodiversity in a general manner, such as the diversity of all living organisms around the globe. However, more generalized definitions can be difficult to apply and work with.

- Explain the focus of the field of **conservation biology**.

- Biodiversity can be defined by multiple levels. Describe the three levels as outlined in the text.

    o

    o

    o

C. Where is most biodiversity?

- To illustrate the distribution of biodiversity across the globe, complete the following chart indicating the degree latitude south of the equator and the corresponding number of species. As an example, the first row is completed for you:

Latitude	Number of land mammals found
Equator	400

- This biodiversity gradient isn't only evident in land mammals.   In what other group(s) or organisms has this trend been observed?

- In evaluating populations around the globe, we can deduce several factors that influence the biodiversity from the equator to the poles. List the three major factors and give a brief explanation of their influence(s).

    1.

    2.

3.

- Define a **biodiversity hotspot**.

- While 25 different specific biodiversity hotspots have been identified, we can group them into three categories. List the categories.

    1.

    2.

    3.

- Why are deep ocean regions considered to be an important part of biodiversity?

D. Island Biogeography Helps Us Understand the Maintenance and Loss of Biodiversity

- Patterns of species diversity can be explained and predicted by the island biogeography theory.
    - The two major components include the _____ effect and the _____ effect.
    - Describe the basic premises for these two components studied by MacArthur and Wilson.

- How did MacArthur and Wilson test their theory?
  Describe the major results.

- Is this theory only applicable to islands? Where else might it be applied? Explain.

II. Extinction Reduces Biodiversity

A. There Are Multiple Causes of Extinction

- The two major categories of extinction include mass extinctions and background extinctions. Describe the differences between the two.

- There may be certain characteristics of a species that cause the species to be at a higher risk for a background extinction.

    1. What role does the *geographic range* play?

    2. What role does the *local population size* play?

    3. What role does *habitat tolerance* play?

B. We Are in the Midst of a Mass Extinction

- Explain how the current rates of extinction compare to the historic rates.

- Describe the various roles humans have in these extinctions (describe, specifically, at least two).

III. Human Interference Generally Reduces Biodiversity

A. Some Ecosystem Disturbances are Reversible, Others Are Not

- Explain the key difference between a reversible ecosystem disturbance and an irreversible ecosystem disturbance.

B. Disruptions of Ecosystems Can Be Disastrous

  1. Introductions of exotic species

- Define an **exotic species**.

  o   Describe two specific reasons why exotic species can be a problem.

- Give an example of an exotic species that entered a new habitat accidentally and an example of a species that was introduced intentionally (and explain why).

  2. Acid rain and the burning of fossil fuels

- All forms of precipitation—including fog, snow, and sleet—can be acidic.

- The major acids involved in acid precipitation are nitric acid and sulfuric acid. Explain how these acids are formed.

- Acid precipitation that falls to the ground can affect various organisms directly and indirectly. Explain, and give an example of both impacts.

- How can we prevent or reduce the presence of these acids in the environment?

3. Global warming and the greenhouse effect

- Define the greenhouse effect.

- Describe the characteristics of a greenhouse gas and give two examples.

- What causes the levels of greenhouses gases to increase?

- The increasing average temperature has an impact on biological systems and physical environments. Describe some specific examples of these impacts.

- What are some ways to reduce emissions of greenhouse gases?

4. Depletion of the ozone layer

- The chemical formula for ozone is _____. Explain how this molecule impacts us when it is at ground level and also when it is in the upper stratosphere.

- Chemicals that are responsible for depleting ozone in the stratosphere are called _____. What are, or were, some of the sources of these chemicals?

- What is the difference between a general ozone reduction and "ozone holes"?

- Without certain levels of ozone in the stratosphere, short-wavelength ultraviolet light is able to reach the surface of the earth. Describe the two majors concerns we have with elevated levels of UVB light.

    o

    o

5. **Deforestation of tropical rain forests**

- There are many factors that influence the destruction of tropical forests. Briefly describe how each of the following plays a role in this problem.

    o Agriculture

    o Pollution from oil wells

    o Mining

    o Road development

- We know that the environmental impact of this destruction is tremendous and one of the more serious environmental problems of our day. Explain the two main consequences of tropic rainforest deforestation.

    1.

    2.

IV. We Can Develop Strategies for Effective Conservation

A. With Limited Resources, We Must Prioritize Which Species Should Be Preserved

- Although the goal of preserving all biodiversity may not be attainable, outline other strategies for devising conservation goals when attempting to answer the question "who to save?"

- Conservation law in the United States involves a specific policy called the **Endangered Species Act**. What does this act define and how can it be used as a tool to preserve species?

B. There Are Multiple Strategies for Preserving Biodiversity

- Some preservation strategies have focused on an individual species. More comprehensive strategies have recently focused on preserving habitats, communities, and entire ecosystems.

- A related approach, often referred to as _____ **conservation**, is a strategy to conserve habitats and ecological processes.

- In outlining preservation areas, there are some beneficial tactics to employ in the design process.
    - What is the benefit of including a *corridor* in an area of preservation?

    - What is the benefit of including a *buffer zone* in an area of preservation?

- In a continued effort to reach beyond an individual species, some biologists focus on specific groups that may in turn benefit other, larger numbers of species. Define the following:

    1. Flagship species:

    2. Keystone species:

    3. Indicator species:

    4. Umbrella species:

Testing and Applying Your Understanding

Multiple Choice (For more multiple choice questions, visit www.prep-u.com.)

1. In conservation biology, to what does the term "corridor" refer?
    a) the edge of a given patch of habitat
    b) sections of habitat organisms use to transverse between two or more isolated patches of habitat
    c) the path a conservative biologist takes through a habitat to minimize any negative effects on the habitat's biodiversity
    d) the path an organism takes from its home to its food source
    e) None of the above.

2. Which of the following factors is currently the leading cause of extinction?
    a) habitat loss
    b) disease
    c) pollution
    d) introduced species
    e) overexploitation

3. Exotic species are a significant threat because they:
    a) may outcompete or prey upon native species.
    b) generally have no natural predators to keep their population in check.
    c) can cause significant structural and economic damage.
    d) a) and b) only.
    e) all of the above.

4. An umbrella species:
    a) is a species whose removal will severely damage or cause the collapse of its ecosystem.
    b) is a term used for any species with a larger-than-expected effect on its ecosystem.
    c) is a species whose protection results in the protection of many other species as well.
    d) is a non-native species, usually introduced to an area accidentally.
    e) is a popular species that the public tends to rally behind and support.

5. Species particularly vulnerable to extinction include:
    a) species with very little genetic variability.
    b) species with small or declining populations.
    c) species commercially used by humans.
    d) endemic species.
    e) all of the above.

6. Which of the following is NOT considered to be a greenhouse gas?
   a) nitrogen
   b) water vapor
   c) ozone
   d) carbon dioxide
   e) methane

7. Even though there is a carbon cycle, it now appears that carbon dioxide levels are rising around the world. Which of the following best explains why this is true?
   a) Animals give off carbon dioxide during their normal metabolism.
   b) As the atmosphere heats up it can contain more carbon dioxide.
   c) The burning of fossil fuels releases more carbon dioxide into the atmosphere.
   d) More carbon dioxide is being given off by ocean waters as they heat up.
   e) The destruction of coral reefs leads to increased levels of carbon dioxide.

8. The mongooses in Hawaii and the cane toads in Australia are both examples of biological control projects gone wrong. What failure was common to both of these projects?
   a) The introduced predator had an activity pattern that was different from the intended prey so the introduction did not result in predation upon the desired species.
   b) The introduced species' reproductive rate was too low to keep up with the population growth of the desired prey species, so no control occurred.
   c) The introduced species was too small to eat adults of the desired prey species.
   d) The introduced species could not eat the intended prey because it was too large, so it started eating native species in the proper size range.
   e) The introduced species successfully controlled the pest that it was introduced to eat and then started to eat native species.

9. Biodiversity hotspots are defined by which two criteria?
   a) species endemism and degree of threat
   b) species richness and size
   c) ecological diversity and species diversity
   d) size and distance from nearest alternative hotspot
   e) species richness and ecosystem integrity

10. In a dynamic equilibrium:
    a) both the number and composition of species on an island remains the same.
    b) the composition of species on an island changes, but the number of species present remains the same.
    c) extinction is the sole determinant of the number of species present on an island.
    d) both the number and composition of species on an island is changing.
    e) immigration rates to an island are reduced to zero.

Short Answer

1. Explain whether there is a consensus on the definition of biodiversity.

2. What is the difference between an endemic species and an exotic species? Could they ever be one in the same? Explain.

3. The theory of island biogeography includes the possibility to predict the number of species on an island while acknowledging the species present will change over time. How do you explain the number of species of birds remaining the same while knowing new species have joined the island?

4. The text briefly explained the procedures MacArthur and Wilson utilized in testing their theory. Outline a few of the strengths and weaknesses of this study design.

5. The dodo bird is one of the more famous species that have been lost to extinction. They were stout, with heavy bodies and short stubby wings. They thrived on an island in the Indian Ocean until settlers arrived clearing land for settlement and bringing along their dogs, cats, and livestock, which roamed freely. It is estimated they were extinct about 80 years after they were first discovered. Explain all the possible factors related to the risk of background extinction that influenced this species extinction.

6. What can or should be done to avoid introduction of exotic species?

7. Why might greenhouses gases be considered "good" at certain levels?

8. This chapter reviewed many major areas of human interference and disruption of biodiversity. Some feel that certain human actions or influences are more detrimental than others. There many never been a consensus on "the worst," but explain (with supporting evidence from the text) what you feel is one of the most significant human interferences today.

1. Scientific Thinking | Your best pathway to understanding the world.

SCIENCE (p. 2)
A body of knowledge based on observation, description, experimentation, and explanation of natural phenomena. [Lat., scientia, knowledge]

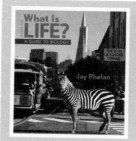

BIOLOGY (p. 4)
The study of living things. [Gk., bios, life + logos, discourse]

SCIENTIFIC LITERACY (p. 4)
A general, fact-based understanding of the basics of biology and other sciences, the scientific method, and the social, political, and legal implications of scientific information.

BIOLOGICAL LITERACY (p. 4)
The ability to use scientific inquiry to think creatively about problems with a biological component; to communicate these thoughts to others; and to integrate these ideas into one's decision-making.

SUPERSTITION (p. 5)
The irrational belief that actions not related by logic to a course of events can influence an outcome.

SCIENTIFIC METHOD (p. 6)
A process of examination and discovery of natural phenomena that involves making observations, constructing hypotheses, testing predictions, experimenting, and drawing conclusions and revising them as necessary.

EMPIRICAL (p. 6)
Describes knowledge that is based on experience and observations that are rational, testable, and repeatable. [Gk., empeiria, experience]

HYPOTHESIS (PL. HYPOTHESES) (p. 9)
A proposed explanation for an observed phenomenon. [Gk., hypothesis, a proposal]

NULL HYPOTHESIS (p. 10)
A hypothesis that proposes a lack of relationship between two factors. [Lat., nullus, none]

CRITICAL EXPERIMENT (p. 12)
An experiment that makes it possible to determine decisively between alternative hypotheses.

PLACEBO (p. 13)
An inactive substance used in controlled experiments to test the effectiveness of another substance; the treatment group receives the substance being tested, the control group receives the placebo. [Lat., placebo, I shall please]

THEORY (p. 15)
An explanatory hypothesis for a natural phenomenon that is exceptionally well supported by empirical data. [Gk., theorein, to consider]

TREATMENT (p. 16)
Any condition applied to the subjects of a research study that is not applied to subjects in a control group.

EXPERIMENTAL GROUP (p. 16)
In an experiment, the group of subjects exposed to a particular treatment; also known as the treatment group.

CONTROL GROUP (p. 16)
In an experiment, the group of subjects not exposed to the treatment being studied but otherwise treated identically to the experimental group.

VARIABLES (p. 16)
The characteristics of an experimental system subject to change; for example, time (the duration of treatment) or specific elements of the treatment, such as the substance or procedure administered or the temperature at which it takes place. [Lat., variare, to vary]

PLACEBO EFFECT (p. 18)
A frequently observed and poorly understood phenomenon in which there is a positive response to treatment with an inactive substance.

BLIND EXPERIMENTAL DESIGN (p. 18)
An experimental design in which the subjects do not know what treatment (if any) they are receiving.

DOUBLE-BLIND EXPERIMENTAL DESIGN (p. 19)
An experimental design in which neither the subjects nor the experimenters know what treatment (if any) individual subjects are receiving.

RANDOMIZED (p. 19)
Describes a manner of choosing subjects and assigning them to groups on the basis of chance, that is, randomly.

INDEPENDENT VARIABLE (p. 23)
A measurable entity that is available at the start of a process being observed and whose value can be changed as required; generally represented on the x-axis in a graph.

DEPENDENT VARIABLE (p. 23)
A measurable entity that is created by the process being observed and whose value cannot be controlled; generally represented on the y-axis in a graph and expected to change in response to a change in the independent variable.

STATISTICS (p. 24)
A set of analytical and mathematical tools designed to further the understanding of numerical data.

↑↑ POSITIVE CORRELATION (p. 25)
A relationship between variables in which they increase (or decrease) together. [Lat., com, with + relatio, report]

PSEUDOSCIENCE (p. 26)
Hypotheses and theories not supported by trustworthy and methodical scientific studies. [Gk., pseudes, false + Lat., scientia, knowledge]

ANECDOTAL OBSERVATION (p. 26)
Observation of one or only a few instances of a phenomenon.

2. Chemistry | Raw materials and fuel for our bodies

ELEMENT (p. 38)
A substance that can not be broken down chemically into any other substances; all atoms of an element have the same atomic number. [Lat., elementum, element, or first principle]

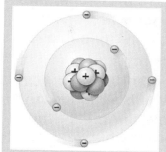

ATOM (p. 39)
A particle of matter than cannot be further subdivided without losing its essential properties. [Gk., atomos, indivisible]

NUCLEUS (p. 39)
The central and most massive part of an atom, usually made up of two types of particles, protons and neutrons, which move about the nucleus. [Lat., nux, nut]

+ Proton   Neutron

PROTON (p. 39)
A positively charged particle in the atomic nucleus; it is identical with the nucleus of the hydrogen atom, which lacks a neutron, and has atomic number 1. [Gk., protos, first]

NEUTRON (p. 39)
An electrically neutral particle in the atomic nucleus.

MASS (p. 39)
The amount of matter in a given sample of a substance.

ELECTRON (p. 39)
A negatively charged particle that moves around the atomic nucleus.

ISOTOPE (p. 40)
Variants of atoms that differ in the number of neutrons they possess; isotopes do not vary in charge, because neutrons have no electrical charge, but the atom's mass changes with the loss or addition of a particle in the nucleus.

6
C
Carbon
12.0107

ATOMIC NUMBER (p. 39)
The number of protons in the nucleus of an atom of a given element.

ATOMIC MASS (p. 39)
The mass of an atom; the combined mass of the protons and neutrons in an atom (the weight of the electrons is negligible).

ISOTOPE (p. 40)
Variants of atoms that differ in the number of neutrons they possess; isotopes do not vary in charge, because neutrons have no electrical charge, but the atom's mass changes with the loss or addition of a particle in the nucleus.

RADIOACTIVE (p. 40)
The property of some elements or isotopes of having a nucleus that breaks down spontaneously, releasing tiny, high-speed particles that carry energy.

PERIODIC TABLE (p. 40)
A tabular display in which all the known chemical elements are arranged in the order of their atomic number and on the basis of other aspects of their atomic structures.

ION (p. 42)
An atom that carries an electrical charge, positive or negative, 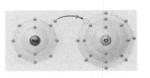 because it has either gained or lost an electron or electrons from its normal, stable configuration. [Gk., ion, going]

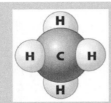

MOLECULE (p. 43)
A group of atoms of the same or different elements held together by bonds; a molecule of a compound of different elements is the smallest unit of that compound that retains its characteristics. [Lat., dim. of moles, mass]

BOND ENERGY (p. 43)
The strength of a bond between two atoms, defined as the energy required to break the bond.

COVALENT BOND (p. 43)
A strong bond formed when two atoms share electrons; the simplest example is the H2 molecule, in which each  of the two atoms in the molecule shares its lone electron with the other atom. [Lat., con-, together + valere, to be strong]

DOUBLE BOND (p. 44)
The sharing of two electrons between two atoms; for example, the most common form of oxygen is the O2 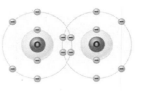 molecule, in which two electrons from each of the two atoms of oxygen are shared.

IONIC BOND (p. 44)
A bond created by the transfer of one or more electrons 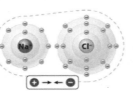 from one atom to another; the resulting atoms, now called ions, are charged oppositely and so attract each other to form a compound.

COMPOUND (p. 44)
A substance composed of atoms of different 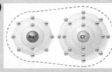 elements in specific ratios, held together by ionic bonds. [Lat., componere, to put together]

HYDROGEN BOND (p. 44)

A type of weak chemical bond formed between the slightly positively charged hydrogen atoms of one molecule and the slightly negatively charged atoms of another (often oxygen or nitrogen atoms); hydrogen bonds are important in building multi-atom molecules, such as complex proteins.

pH (p. 50)

A logarithmic scale that measures the concentration of hydrogen ions ($H+$) in a solution, with decreasing values indicating increasing acidity; water, in which the concentration of hydrogen ions ($H+$) equals the concentration of hydroxyl ions ($OH-$), is pH =7, the midpoint of the scale. [abbreviation for "power of hydrogen"]

ACID (p. 50)

Any fluid with a pH below 7.0, that is, with more $H+$ ions than $OH-$ ions. [Lat., acidus, sour]

BASE (p. 50)

Any fluid with a pH above 7.0, that is, with more $OH-$ ions than $H+$ ions.

MACROMOLECULE (p. 52)

A large molecule, made up of smaller building blocks or subunits; four types of biological macro-molecules are carbohydrates, lipids, proteins, and nucleic acids. [Gk., macros, large + Lat., dim. of moles, mass]

MONOSACCHARIDES (OR SIMPLE SUGARS) (p. 52)

The simplest carbohydrates and the building blocks of more complex carbohydrates; they cannot be broken down into other monosaccharides; examples are glucose, fructose, and galactose; also known as simple sugars. [Gk., monos, single +sakcharon, sugar]

BUFFER (p. 51)

A chemical that can quickly absorb excess $H+$ ions in a solution (preventing it from becoming too acidic) or quickly release $H+$ ions to counteract increases in $OH-$ concentration.

CARBOHYDRATE (p. 52)

One of the four types of macro-molecule, containing mostly carbon, hydrogen, and oxygen; carbohydrates are the primary fuel for cellular activity and form much of the cell structure in all life forms. [Lat. carbo, charcoal + hydro-, pertaining to water]

GLYCOGEN (p. 53)

A complex carbohydrate consisting of stored glucose molecules linked to form a large web, which breaks down to release glucose when it is needed for energy. [Gk., glykys, sweet + Gk., genos, race, descent]

DISACCHARIDES (p. 54)

Carbohydrates formed by the union of two  simple sugars, such as sucrose (table sugar) and lactose (the sugar found in milk). [Gk., di-, two + sakcharon, sugar]

POLYSACCHARIDES (p. 54)

Complex carbohydrates formed by the union of many simple sugars. [Gk., polys, many + sakcharon, sugar]

STARCH (p. 55)

A complex polysaccharide carbohydrate consisting of a large number of monosaccharides linked in line; in plants, starch is the primary form of energy storage.

CHITIN (p. 56)

(pron. KITE-in) A complex carbohydrate, indigestible by humans, that forms the rigid outer skeleton of most insects and crustaceans. [Gk., chiton, undershirt]

CELLULOSE (p. 56)

A complex carbohydrate, indigestible by humans, that serves as the structural material for a huge variety of plant structures. It is the single most prevalent organic compound on earth. [Lat., cellula, dim. of cella, room]

LIPID (p. 57)

One of four types of macromolecules, lipids are insoluble in water and greasy to the touch; they are important in energy storage and insulation (fats), membrane formation (phospholipids), and regulating growth and development (sterols). [Gk., lipos, fat]

HYDROPHOBIC (p. 57)

Repelled by water, as, for example, nonpolar molecules that tend to minimize contact with water. [Lat., hydro-, pertaining to water, Gk., phobos, fearing]

HYDROPHILIC (p. 57)

Attracted to water, as, for example, polar molecules that readily form hydrogen bonds with water. [Lat., hydro-, pertaining to water; Gk., philios, loving]

GLYCEROL (p. 58)

A small molecule that forms the head region of a triglyceride fat molecule. [Gk., glykys, sweet + -ol, chemical suffix for an alcohol]

FATTY ACID (p. 58)

A long hydrocarbon (a chain of carbon-hydrogen molecules); fatty acids form the tail region of triglyceride fat molecules.

TRIGLYCERIDE (p. 58)

A fat having three fatty acids linked to the glycerol molecule. [Gk., tri- three + glykys, sweet]

SATURATED FAT (p. 60)
A fat in which each carbon in the hydrocarbon chain forming the tail region of the fat molecule is bound to two hydrogen atoms; saturated fats are solid at room temperature.

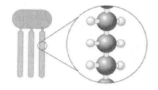

STEROL (p. 61)
A lipid important in regulating growth and development; the sterols include cholesterol and the steroid hormones testosterone and estrogen, and all are modifications of a basic structure of four interlinked rings of carbon atoms. [Gk., stereos, solid + -ol, chemical suffix for an alcohol]

UNSATURATED FAT (p. 60)
A fat in which at least one carbon in the hydrocarbon chain forming the tail region of the fat molecule is bound to only one hydrogen atom; unsaturated fats are liquid at room temperature.

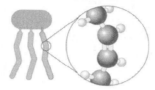

CHOLESTEROL (p. 61)
One of the sterols, a group of lipids important in regulating growth and development; an important component of most cell membranes, helping the membrane maintain its flexibility. [Gk., chole, bile + stereos, solid + -ol, chemical suffix for an alcohol]

TRANS FAT (p. 60)
An unsaturated fat that has been partially hydrogenated (meaning that hydrogen atoms have been added to make the fat more saturated and to improve a food's taste, texture, and shelf-life). The added hydrogen atoms are in a trans orientation, which differs from the cis ("near") orientation of hydrogen atoms in the unsaturated fat. [Lat., trans, on the other side of]

PHOSPHOLIPIDS (p. 62)
A lipid that is the major component of the plasma membrane; phospholipids are structurally similar to fats, but contain a phosphorus atom and have two, not three, fatty acid chains.

WAX (p. 62)
A lipid similar in structure to fats but with only one long-chain fatty acid linked to the glycerol head of the molecule; because the fatty acid chain is highly nonpolar, waxes are strongly hydrophobic.

PROTEIN (p. 63)
One of the four types of biological macromolecules; constructed of unique combinations of 20 amino acids that result in unique structures and chemical behavior. Proteins are the chief building blocks of tissues in most organisms. [Gk., proteion, of the first quality]

ENZYME (p. 63)
A protein that initiates and accelerates a chemical reaction in a living organism; enzymatic proteins take part in chemical reactions on the inside and outside surfaces of the plasma membrane. [Gk., en, in + zyme, leaven]

AMINO ACID (p. 63)
One of 20 molecules built of an amino group, a carboxyl group, and a unique side chain. Proteins are constructed of combinations of amino acids linked together.

CARBOXYL GROUP (p. 64)
A functional group characterized by a carbon atom double-bonded to one oxygen atom and single-bonded to another oxygen atom. Amino acids are made up of an amino group, a carboxyl group, and a side chain.

PEPTIDE BOND (p. 66)
A bond in which the amino group of one amino acid is bonded to the carboxyl group of another; two amino acids so joined form a dipeptide, several amino acids so joined form a polypeptide. [Gk., peptikos, able to digest]

AMINO GROUP (p. 64)
A nitrogen atom bonded to three hydrogen atoms.

PRIMARY STRUCTURE (p. 66)
The sequence of amino acids in a polypeptide chain.

SECONDARY STRUCTURE (p. 66)
The corkscrew-like twists or folds formed by hydrogen bonds between amino acids in a polypeptide chain.

TERTIARY STRUCTURE (p. 66)
The unique and complex three-dimensional shape formed by multiple twists of the secondary structure of a protein as amino acids come together and form hydrogen bonds or covalent sulfur-sulfur bonds. [Lat., tertius, third]

QUATERNARY STRUCTURE (p. 66)
Two or more polypeptide chains bonded together in a single protein; an example is hemoglobin. [Lat., quaterni, four each]

ACTIVE SITE (p. 68)
The part of an enzyme to which reactants (or substrates) bind and undergo a chemical reaction.

DENATURATION (p. 67)
The disruption of protein folding, in which secondary and tertiary structures are lost, caused by exposure to extreme conditions in the environment such as heat or extreme pH.

SUBSTRATE (p. 68)
The molecule on which an enzyme acts. The active site on the enzyme binds to the substrate, initiating a chemical reaction; for example, the active site on the enzyme lactase binds to the substrate lactose, breaking it down into the two simple sugars glucose and galactose. [Lat., sub-, under + stratus, spread]

ACTIVATION ENERGY (p. 68)
The minimum energy needed to initiate a chemical reaction (regardless of whether the reaction releases or consumes energy).

INHIBITOR (p. 69)
A chemical within a cell that binds to an enzyme or substrate molecule and in doing so reduces the enzyme's ability to catalyze a reaction.

COMPETITIVE INHIBITOR (p. 69)
A chemical within a cell that binds to the active site of an enzyme, blocking substrate molecules from the site and thereby reducing the enzyme's ability to catalyze a reaction.

NONCOMPETITIVE INHIBITOR (p. 69)
A chemical within a cell that binds to part of an enzyme away from the active site, but alters the enzyme's shape so as to alter the active site, thereby reducing or blocking the enzyme's ability to bind with substrate.

ACTIVATOR (p. 70)
A chemical within a cell that binds to an enzyme, altering the enzyme's shape

NUCLEIC ACID (p. 71)
One of the four types of biological macromolecules; the nucleic acids DNA and RNA store genetic information in unique sequences of nucleotides.

NUCLEOTIDE (p. 71)
A molecule containing a phosphate group, a sugar molecule, and a nitrogen-containing molecule. Nucleotides are the individual units that together, in a unique sequence, constitute a nucleic acid.

DEOXYRIBONUCLEIC ACID (p. 71)
A nucleic acid, DNA carries information about the production of particular proteins in the sequences of its nucleotide bases.

RIBONUCLEIC ACID (p. 71)
A nucleic acid, RNA serves as a middleman in the process of converting genetic information in DNA into protein; messenger RNA (mRNA) takes instructions for production of a given protein from DNA to another part of the cell, whereas transfer RNA (tvRNA) interprets the mRNA code and directs the construction of the protein from its constituent amino acids.

BASE (OF DNA) (p. 71)
One of the nitrogen-containing side-chain molecules attached to a sugar molecule in the sugar-phosphate backbone of DNA and RNA. The four bases in DNA are adenine (A), thymine (T), guanine (G), and cytosine (C); the four bases in RNA are adenine (A), uracil (U), guanine (G), and cytosine (C). The information in a molecule of DNA and RNA is determined by its sequence of bases.

DOUBLE-HELIX (p. 72)
The spiraling ladder-like structure of DNA composed of two strands of nucleotides; the bases protruding from eachstrand like "half-rungs" meet in the center and bind to each other (via hydrogen bonds), holding the ladder together. [Gk., heligmos, wrapping]

3. Cells

The smallest part of you

VISUAL LEARNING GLOSSARY: Key Terms in the order you see them in the text.

CELL (p. 82)
The smallest unit of life that can function independently; a three-dimensional structure, surrounded by a membrane and, in the case of prokaryotes and most plants, a cell wall, in which many of the essential chemical reactions of the life of an organism take place. [Lat., cella, room]

EUKARYOTE (p. 84)
An organism composed of eukaryotic cells. [Gk., eu, good + karyon, nut, kernel]

CELL THEORY (p. 84)
A unifying and universally accepted theory in biology that holds that all living organisms are made up of one or more cells, and that all cells arise from other, preexisting cells.

PROKARYOTIC CELL (p. 84)
A cell bound by a plasma membrane enclosing the cell contents (cytoplasm, DNA, and ribosomes); there is no nucleus or other organelles.

EUKARYOTIC CELL (p. 84)
A cell with a membrane-surrounded nucleus containing DNA, membrane-surrounded organelles, and internal structures organized into compartments.

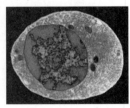

PROKARYOTE (p. 84)
An organism consisting of a prokaryotic cell (all prokaryotes are one-celled organisms). [Gk., pro, before + karyon, nut, kernel]

PLASMA MEMBRANE (p. 84)
A complex, thin, two-layered membrane that encloses the cytoplasm of the cell, holding the contents in place and regulating what enters and leaves the cell; also called the cell membrane. [Gk., plasma, anything molded]

CYTOPLASM (p. 84)
The jelly-like fluid that fills the inside of the cell; in eukaryotes, the cytoplasm contains the organelles. [Gk., kytos, container + plasma, anything molded]

RIBOSOMES (p. 84)
Granular bodies in the cytoplasm, released from their initial positions on the rough endoplasmic reticulum, that copy the information in segments of DNA to provide instruction for the construction of proteins.

CELL WALL (p. 85)
A rigid structure, outside the cell membrane, that protects and gives shape to the cell; found in many prokaryotes and plants.

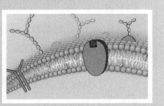

VLG - 10

FLAGELLUM (p. 85)
Long, thin, whip-like projection from the cell body of a prokaryote that aids in cell movement through the medium in which the organism lives; in animals, the only cell with a flagellum is the sperm cell. [Lat., flagellum, whip]

PILUS (p. 85)
A thin, hair-like projection that helps a prokaryote attach to surfaces. [Lat., pilus, a single hair]

NUCLEUS (p. 86)
A membrane-enclosed structure in eukaryotic cells that contains the organism's genetic information as linear strands of DNA in the form of chromosomes. [Lat., nucleus, dim. of nux, nut]

ORGANELLES (p. 86)
Specialized structures in the cytoplasm of eukaryotic cells with specific functions, such as the rough and smooth endoplasmic reticulum, Golgi apparatus, and mitochondria. [Gk., organon, tool]

CHLOROPLAST (p. 88)
The organelle in plant cells in which photosynthesis occurs. [Gk., chloros, pale green + plastos, formed]

MITOCHONDRION (p. 88)
The organelle in eukaryotic cells that converts the energy stored in food in the chemical bonds of carbohydrate, fat, and protein molecules into a form usable by the cell for all its functions and activities. [Gk., mitos, thread + chondros, cartilage]

ENDOSYMBIOSIS THEORY (p. 86)
Theory of the origin of eukaryotes that holds that, in the past, two different types of prokaryotes engaged in a close partnership and eventually one, capable of performing photosynthesis, was subsumed into the other, a larger prokaryote. The smaller prokaryote made some of its photosynthetic energy available to the host and, over time, the two became symbiotic and eventually became a single more complex organism in which the smaller prokaryote had evolved into the chloroplast of the new organism. A similar scenario can be developed for the evolution of mitochondria. [Gk., endon, within + symbios, living together]

INVAGINATION (p. 88)
The folding in of a membrane or layer of tissue so that an outer surface becomes an inner surface. [Lat., in, in + vagina, sheath]

PHOSPHOLIPIDS (p. 89)
A lipid that is the major component of the plasma membrane; phospholipids are structurally similar to fats, but contain a phosphorus atom and have two, not three, fatty acid chains.

GLYCEROL (p. 89)
A small molecule that forms the head region of a triglyceride fat molecule. [Gk., glykys, sweet + -ol, chemical suffix for an alcohol]

POLAR (p. 89)
Having an electrical charge.

NONPOLAR (p. 89)
Electrically uncharged.

HYDROPHOBIC (p. 89)
Repelled by water, as, for example, nonpolar molecules that tend to minimize contact with water. [Lat., hydro-, pertaining to water, Gk., phobos, fearing]

HYDROPHILIC (p. 89)
Attracted to water, as, for example, polar molecules that readily form hydrogen bonds with water. [Lat., hydro-, pertaining to water; Gk., philios, loving]

PHOSPHOLIPID BILAYER (p. 90)
The structure of the plasma membrane; two layers of phospholipids, arranged tail to tail (the tails are hydrophobic and so avoid contact with water), with the hydrophilic head regions facing the watery extracellular fluid and intracellular fluid.

TRANSMEMBRANE PROTEIN (p. 91)
A protein that can penetrate the phospholipid bilayer of a cell's plasma membrane.

SURFACE PROTEIN (p. 91)
A protein that resides primarily on the inner or outer surface of the phospholipid bilayer that constitutes the plasma membrane of the cell.

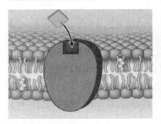

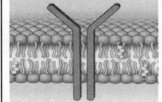

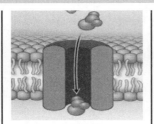

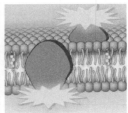

RECEPTOR PROTEIN (p. 91)
A protein in the plasma membrane that binds to specific chemicals in the cell's external environment to regulate processes within the cell; for example, cells in the heart have receptor proteins that bind to adrenaline.

RECOGNITION PROTEIN (p. 92)
A protein in the plasma membrane that provides a "fingerprint" on the outside-facing surface of the cell, making it recognizable to other cells. Recognition proteins make it possible for the immune system to distinguish the body's own cells from invaders that may produce infection, and also help cells bind to other cells or molecules.

TRANSPORT PROTEIN (p. 92)
A transmembrane protein that provides a channel or passageway through which large or strongly charged molecules can pass. Transport proteins are of a number of shapes and sizes, making possible the transport of a wide variety of molecules.

ENZYMATIC PROTEIN (ENZYME) (p. 92)
A protein that initiates and accelerates a chemical reaction in a living organism; enzymatic proteins take part in chemical reactions on the inside and outside surfaces of the plasma membrane. [Gk., en, in + zyme, leaven]

CHOLESTEROL (p. 92)
One of the sterols, a group of lipids important in regulating growth and development; an important component of most cell membranes, helping the membrane maintain its flexibility. [Gk., chole, bile + stereos, solid + -ol, chemical suffix for an alcohol]

FLUID MOSAIC (p. 92)
A term that describes the structure of the plasma membrane, which is made up of several different types of molecules, many of which are not fixed in place but float, held in proper orientation by hydrophilic and hydrophobic forces.

PASSIVE TRANSPORT (p. 97)
Molecular movement that occurs spontaneously, without the input of energy; the two types of passive transport are diffusion and osmosis.

DIFFUSION (p. 97)
Passive transport in which a particle (the solute) is dissolved in a gas or liquid (the solvent) and moves from an area of higher solute concentration to an area of lower solute concentration. [Lat., diffundere, to pour in different directions]

SOLUTE (p. 97)
A substance that is dissolved in a gas or liquid; in a solution of water and sugar, sugar is the solute. [Lat., solvere, to loosen]

SOLVENT (p. 97)
The gas or liquid in which a substance is dissolved; in a solution of water and sugar, water is the solvent. [Lat., solvere, to loosen]

SIMPLE DIFFUSION (p. 97)
Diffusion of molecules directly through the phospholipid bilayer of the plasma membrane that takes place without the assistance of other molecules; oxygen and carbon dioxide, because they are small and carry no charge that would cause them to be repelled by the middle layer of the membrane, can pass through the membrane in this way.

FACILITATED DIFFUSION (p. 98)
Diffusion of molecules through the phospholipid bilayer of the plasma membrane that takes place through a transport protein (a "carrier molecule") embedded in the membrane. Molecules that require the assistance of a carrier molecule are those that are too big to cross the membrane directly or are electrically charged and would be repelled by the middle layer of the membrane.

OSMOSIS (p. 99)
A type of passive transport in which water molecules move across a membrane, such as the plasma membrane of a cell; the direction of osmosis is determined by the relative concentrations of all solutes on either side of the membrane. [Gk., osmos, thrust]

TONICITY (p. 99)
For a cell in solution, a measure of the concentration of solutes outside the cell relative to that inside the cell. [Gk., tonos, tension]

=

ISOTONIC (p. 99) Refers to solutions with equal concentrations of solutes. [Gk., isos, equal to + tonos, tension]

<

HYPOTONIC (p. 99) Of two solutions, that with a lower concentration of solutes. [Gk., hypo, under + tonos, tension]

>

HYPERTONIC (p. 99) Of two solutions, that with a higher concentration of solutes. [Gk., hyper, above + tonos, tension]

ACTIVE TRANSPORT (p. 101) Molecular movement that depends on the input of energy, which is necessary when the molecules to be moved are large or are being moved against their concentration gradient.

PRIMARY ACTIVE TRANSPORT (p. 101) Active transport using energy released directly from ATP.

SECONDARY ACTIVE TRANSPORT (p. 101) Active transport in which there is no direct involvement of ATP (adenosine triphosphate); the transport protein simultaneously moves one molecule against its concentration gradient while letting another flow down its concentration gradient.

ENDOCYTOSIS (p. 102) A cellular process in which large particles, solid or dissolved, outside the cell are surrounded by a fold of the plasma membrane, which pinches off, forming a vesicle, and the enclosed particle now moves into the cell. The three types of endocytosis are phagocytosis, pinocytosis, and receptor-mediated endocytosis. [Gk., endon, within + kytos, container]

EXOCYTOSIS (p. 102) A cellular process in which particles within the cell, solid or dissolved, are enclosed in a vesicle and transported to the plasma membrane, where the membrane of the vesicle merges with the plasma membrane and the material in the vesicle is expelled to the extracellular fluid for use throughout the body. [Gk., ex, out of + kytos, container]

VESICLE (p. 102) A small, membrane-enclosed sac within a cell. [Lat., vesicula, dim. of vesica, bladder]

PHAGOCYTOSIS (p. 102) One of the three types of endocytosis, in which relatively large solid particles are engulfed by the plasma membrane, a vesicle is formed, and the particle is moved into the cell.

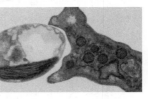

PINOCYTOSIS (p. 102) One of the three types of endocytosis, in which dissolved particles and liquids are engulfed by the plasma membrane, a vesicle is formed, and the material is moved into the cell. The vesicles formed in pinocytosis are generally much smaller than those formed in phagocytosis. [Gk., pinein, to drink + kytos, container]

RECEPTOR-MEDIATED ENDOCYTOSIS (p. 102) One of the three types of endocytosis, in which receptors on the surface of a cell bind to specific molecules; the plasma membrane then engulfs both molecule and receptor and draws them into the cell.

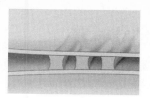

TIGHT JUNCTION (p. 105)
A continuous, water-tight connection between adjacent animal cells. Tight junctions are particularly important in the small intestine, where digestion occurs, to ensure that nutrients do not leak between cells into the body cavity and so become lost as a source of energy.

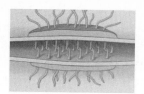

DESMOSOMES (p. 105)
Irregularly spaced connections between adjacent animal cells that, in the manner of Velcro, hold cells together by attachments but are not water-tight. They provide mechanical strength and are found in muscle tissue and in much of the tissue that lines the cavities of animal bodies. [Gk., desmos, bond + soma, body]

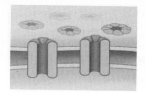

GAP JUNCTION (p. 105)
A junction between adjacent animal cells in the form of a pore in each of the plasma membranes surrounded by a protein that links the two cells and acts like a channel between them, allowing materials to pass between the cells.

NUCLEAR MEMBRANE (p. 107)
A membrane that surrounds the nucleus of a cell, separating it from the cytoplasm, consisting of two bilayers and perforated by pores enclosed in embedded proteins that allow the passage of large molecules from nucleus to cytoplasm and from cytoplasm to nucleus; also called the nuclear envelope.

CHROMATIN (p. 108)
A mass of long, thin fibers consisting of DNA and proteins in the nucleus of the cell. [Gk., chroma, color]

NUCLEOLUS (p. 108)
An area near the center of the nucleus where subunits of the ribosomes are assembled. [Lat., nucleolus, dim. of nucleus, kernel, small nut]

CYTOSKELETON (p. 109)
A network of protein structures in th cytoplasm of eukaryotes (and, to a lesser extent, prokaryotes) that serves as scaffolding, adding support and, in some cases, giving animal cells of different types their characteristic shapes. The cytoskeleton serves as tracks guiding the intercellular traffic flow and, because it is flexible and can generate force, gives cells some ability to control their movement.

MICROTUBULES (p. 109)

One of three types of protein fibers (the others are intermediate filaments and micro-filaments) that make up the eukaryotic cytoskeleton, providing it with structure and shape, These are the thickest elements in the cytoskeleton; they resemble rigid, hollow tubes, functioning as tracks to which mole-cules and organelles within the cell may attach and be moved along; also help pull chromo-somes apart during cell division.

INTERMEDIATE FILAMENTS (p. 109)

One of three types of protein fibers (the others are microtubules and microfilaments) that make up the eukary-otic cytoskeleton, providing it with structure and shape; a durable, rope-like system of numerous differentoverlapping proteins.

MICROFILAMENTS (p. 109)

One of three types of protein fibers (the others are intermediate filaments and microtubules) that make up the eukaryotic cytoskeleton, providing it with structure and shape, These are the thinnest elements in the cytoskel-eton; long, solid, rod-like fibers that help generate forces, including those important in cell contrac-tion and cell division.

CILIA (p. 110)

Short projections from the cell surface, often occurring in large numbers on a single cell, that beat against the intercellular fluid to move the fluid past the cell. [Lat., cilium, eyelid]

INTERMEMBRANE SPACE (p. 112)

In a mitochondrion, the region between the inner and outer membranes. [Lat., inter, between + membrana, a thin skin]

LYSOSOME (p. 112)

A round, membrane-enclosed, enzyme- and acid-filled vesicle in the cell that digests and recycles cellular waste products and consumed material. [Gk., lysis, releasing + soma, body]

MITOCHONDRION (p. 110)

The organelle in eukaryotic cells that converts the energy stored in food in the chemical bonds of carbohydrate, fat, and protein molecules into a form usable by the cell for all its functions and activities. [Gk., mitos, thread + chondros, cartilage]

MATRIX (MITOCHONDRIAL) (p. 112)

The space within the inner membrane, where the carriers NADH and FADH2 begin the electron transport chain by carrying high-energy electrons to molecules embedded in the inner membrane.

ENDOMEMBRANE SYSTEM (p. 114)

A system of organelles (the rough endoplasmic reticulum, the smooth endoplasmic reticulum, and the Golgi apparatus) that surrounds the nucleus; it produces and modifies nec-essary molecules, breaks down toxic chemicals and cellular by-products, and is thus responsible for many of the fundamental functions of the cell. [Gk., endon, within + Lat., membrana, a thin skin]

ROUGH ENDOPLASMIC RETICULUM (p. 114)

An organelle, part of the endomembrane system, structurally like a series of interconnected, flattened sacs connected to the nuclear envelope; called "rough" because its surface is studded with ribosomes. [Gk., endon, within + plasma, anything molded; Lat., reticulum, dim. of rete, net]

GOLGI APPARATUS (p. 116)

(pron. GOHL-jee) An organelle, part of the endomembrane system, structurally like a flattened stack of unconnected membranes, each known as a Golgi body. The Golgi apparatus processes molecules synthesized in the cell and packages those molecules that are destined for use elsewhere in the body. [From the name of the discoverer, Camillo Golgi, 1843–1926]

VACUOLE (CENTRAL) (p. 119)

A membrane-enclosed, fluid-filled, multipurpose organelle prominent in most plant cells (but also present in some protists, fungi, and animals); functions vary but can include storing nutrients, retaining and degrading waste products, accumulating poisonous materials, containing pigments, and providing physical support. [Lat., vacuus, empty]

STROMA (p. 120)

In the leaf of a green plant, the fluid in the inner compartment of a chloroplast, which contains DNA and protein-making machinery. [Gk., stroma, bed]

SMOOTH ENDOPLASMIC RETICULUM (p. 114)

An organelle, part of the endomembrane system, structurally like a series of branched tubes; called "smooth" because its surface has no ribosomes. Smooth endoplasmic reticulum synthesizes lipids such as fatty acids, phospholipids, and steroids. [Gk., endon, within + plasma, anything molded; Lat., reticulum, dim. of rete, net]

PLASMODESMATA (p. 119)

In plants, microscopic tubelike channels connecting the cells and enabling communication and transport between them. [Gk., plassein, to mold + desmos, bond]

TURGOR PRESSURE (p. 120)

In plants, the pressure of the contents of the cell against the cell wall, which is maintained by osmosis as water rushes into the cell when it contains high concentrations of dissolved substances. Turgor pressure allows non-woody plants to stand upright, and its loss causes wilting. [Lat., turgere, to swell]

THYLAKOIDS (p. 120)

Interconnected membranous structures in the stroma of a chloroplast, where light energy is collected and the conversion of light energy to chemical energy in photosynthesis takes place. [Gk., thylakis, dim. of thylakos, bag]

4. Energy | From the sun to you in just two steps

BIOFUELS (p. 130)
Fuels produced from plant and animal products.

FOSSIL FUELS (p. 130)
Fuels produced from the decayed remains of ancient plants and animals; include oil, natural gas, and coal.

PHOTOSYNTHESIS (p. 131)
The process by which some organisms are able to capture energy from the sun and store it in the chemical bonds of sugars and other molecules that the plants produce. [Gk., phos, light+ syn, together with + tithenai, to place or put]

CELLULAR RESPIRATION (p. 131)
The process by which all living organisms extract energy stored in the chemical bonds of molecules and use it for fuel for their life processes.

KINETIC ENERGY (p. 132)
The energy of moving objects, such as legs pushing the pedals of a bicycle or wings beating against the air. [Gk., kinesis, motion]

ENERGY (p. 132)
The capacity to do work, which is the moving of matter against an opposing force. [Gk., energeia, activity]

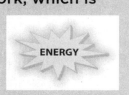

POTENTIAL ENERGY (p. 132)
Stored energy; the capacity to do work that results from an object's location or position, as in the case of water held behind a dam. [Lat., potentia, power]

CHEMICAL ENERGY (p. 133)
A type of potential energy in which energy is stored in chemical bonds between atoms or molecules.

THERMODYNAMICS (p. 132)
The study of the transformation of energy from one type to another, such as from potential energy to kinetic energy.
[Gk., thermÐ, heat + dynamis, power]

FIRST LAW OF THERMODYNAMICS (p. 132)
A physical law that states that energy cannot be created or destroyed; it can only change from one form to another.

 SECOND LAW OF THERMODYNAMICS (p. 135)
A physical law that states that every conversion of energy is not perfectly efficient and invariably includes the transformation of some energy into heat.

ADENOSINE TRIPHOSPHATE (ATP) (p. 135)
A molecule that temporarily stores energy for cellular activity in all living  organisms. ATP is composed of an adenine, a sugar molecule, and a chain of three negatively charged phosphate groups.

CHLOROPLAST (p. 139)
The organelle in plant cells in which photosynthesis occurs. [Gk., chloros, pale green + plastos, formed]

STROMA (p. 139)
In the leaf of a green plant, the fluid in the inner compartment of a chloroplast, which contains DNA and protein-making machinery. [Gk., stroma, bed]

THYLAKOIDS (p. 139)
 Interconnected membranous structures in the stroma of a chloroplast, where light energy is collected and the conversion of light energy to chemical energy in photosynthesis takes place. [Gk., thylakis, dim. of thylakos, bag]

CHLOROPHYLL (p. 139)
A light-absorbing pigment molecule in chloroplasts. [Gk., chloros, pale green + phyllon, leaf]

LIGHT ENERGY (p. 140)
A type of kinetic energy made up of energy packets called photons, which are organized into waves.

PHOTON (p. 140)
The elementary particle that carries the energy of electromagnetic radiation of all wavelengths. [Gk., phos, light]

ELECTROMAGNETIC SPECTRUM (p. 140)
The range of wavelengths that produce electromagnetic radiation, extending (in order of decreasing energy) from high-energy, short-wave, gamma rays and X rays, through ultraviolet light, visible light, and infrared light, to very long, lowenergy, radio waves. [Lat., specere, to look at]

PIGMENT (p. 140)
In photosynthesis, molecules that are able toabsorb the energy of light of specific wavelengths, raising electrons to an excited state in the process. [Lat., pigmentum, paint]

CHLOROPHYLL a (p. 140)

The primary photosynthetic pigment. Chlorophyll a absorbs blue-violet and red light; because it cannot absorb green light and instead reflects those wavelengths, we perceive the reflected light as the color green.

CAROTENOIDS (p. 141)

Pigments that absorb blue-violet and blue-green wavelengths of light and reflect yellow, orange, and red wavelengths of light. [Lat., carota, carrot]

PRIMARY ELECTRON ACCEPTOR (p. 144)

In photosynthesis, a molecule that accepts excited, high-energy electrons from chlorophyll a, beginning the series of electron handoffs known as an electron transport chain.

NADPH (p. 145)

A molecule (nicotinamide adenine dinucleotide phosphate) that is a high-energy electron carrier involved in photosynthesis, which stores energy by accepting high-energy protons. It is formed when the electrons released from the splitting of water are passed to NADP+.

CHLOROPHYLL b (p. 140)

A photosynthetic pigment similar in structure to chlorophyll a. Chlorophyll b absorbs blue and red-orange wavelengths and reflects yellow-green wavelengths.

PHOTOSYSTEMS (p. 143)

Two arrangements of light-absorbing pigments, including chlorophyll, within the chloroplast that capture energy from the sun and transform it first into the energy of excited electrons and ultimately into ATP and high-energy electron carriers such as NADPH. [Gk., phos, light + systema, a whole compounded of parts]

ELECTRON TRANSPORT CHAIN (p. 144)

The path of high-energy electrons from one molecule within a membrane to another, coupled to the pumping of protons across the membrane, creating a concentration gradient that is used to make ATP; occurs in mitochondria and chloroplasts.

CALVIN CYCLE (p. 146)

In photosynthesis, a series of chemical reactions in the stroma of chloroplasts, in which sugar molecules are assembled. [From the name of one of its discoverers, Melvin Calvin, 1911–1997]

RUBISCO (p. 146)

An enzyme (ribulose 1,5-bisphosphate carboxylase/oxygenase), important in photosynthesis, that fixes carbon atoms from CO_2 in the air, attaching them to an organic molecule in the stroma of the chloroplast. This fixation is the first step in the Calvin cycle, in which molecules of sugar are assembled. Rubisco is the most plentiful protein on earth.

STOMATA (Sing. STOMA) (p. 148)

Small pores, usually on the undersides of leaves, that are the primary sites for gas exchange in plants; carbon dioxide (for photosynthesis) enters and oxygen (a by-product of photosynthesis) exits through the stomata. [Gk., stoma, mouth]

CAM PHOTOSYNTHESIS (p. 150)

Energetically expensive photosynthesis in which the stomata are open only at night to admit CO2, which is bound to a holding molecule and released to enter the Calvin cycle to make sugar during the day. In this type of photosynthesis, found in many fleshy, juicy plants of hot, dry areas, water loss is reduced because the stomata are closed during the day.

PYRUVATE (p. 152)

The end product of glycolysis.

KREBS CYCLE (p. 154)

The second step of cellular respiration, in which energy is extracted from sugar molecules as additional molecules of ATP and NADH are formed. [From the name of the discoverer, Hans Adolf Krebs, 1900–1981]

ETHANOL (p. 160)

The end product of fermentation of yeast; the alcohol in beer, wine, and spirits. [contraction of the full chemical name, ethyl alcohol]

C4 PHOTOSYNTHESIS (p. 148)

A method (along with C3 and CAM photosynthesis) by which plants fix carbon dioxide, using the carbon to build sugar; serves as a more effective method than C3 for binding carbon dioxide under low carbon dioxide conditions, such as when plants in warmer climates close their stomata to reduce water loss.

GLYCOLYSIS (p. 152)

In all organisms, the first step in cellular respiration, in which one molecule of glucose is broken into two molecules of pyruvate. For some organisms, glycolysis is the only means of extracting energy from food; for others, including most plants and animals, it is followed by the Krebs cycle and the electron transport chain. [Gk., glykys, sweet + lysis, releasing]

MATRIX (MITOCHONDRIAL) (p. 156)

The space within the inner membrane, where the carriers NADH and FADH2 begin the electron transport chain by carrying high-energy electrons to molecules embedded in the inner membrane.

FERMENTATION (p. 160)

The process by which glycolysis occurs in the absence of oxygen; the electron acceptor is pyruvate (in animals) or acetaldehyde (in yeast) rather than oxygen.

5. DNA, Gene Expression, and Biotechnology

What is the genetic code, and how is it harnessed?

VISUAL LEARNING GLOSSARY: Key Terms in the order you see them in the text.

DEOXYRIBONUCLEIC ACID (DNA) (p. 172)
A nucleic acid, DNA carries information about the production of particular proteins in the sequences of its nucleotide bases.

NUCLEOTIDE (p. 172)
A molecule containing a phosphate group, a sugar molecule, and a nitrogencontaining molecule. Nucleotides are the individual units that together, in a unique sequence, constitute a nucleic acid.

BASE PAIR (p. 173)
Two nucleotides on complementary strands of DNA that form a pair, linked by hydrogen bonds. The pattern of pairing is adenine (A) with thymine (T) and cytosine (C) with guanine (G); the base-paired arrangement forms the "rungs" of the double-helix structure of DNA.

GENOME (p. 175)
The full set of DNA present in an individual organism; also can refer to the full set of DNA present in a species.

GENE (p. 175)
The basic unit of heredity; a sequence of DNA nucleotideson a chromosome that carries the information necessary for making a functional product, usually a protein or an RNA molecule. [Gk., genos, race, descent]

NUCLEIC ACID (p. 172)
One of the four types of biological macromolecules; the nucleic acids DNA and RNA store genetic information in unique sequences of nucleotides.

BASE (p. 172)
One of the nitrogen-containing side-chain molecules attached to a sugar molecule in the sugar-phosphate backbone of DNA and RNA. The four bases in DNA are adenine (A), thymine (T), guanine (G), and cytosine (C); the four bases in RNA are adenine (A), uracil (U), guanine (G), and cytosine (C). The information in a molecule of DNA and RNA is determined by its sequence of bases.

CODE (p. 175)
In genetics, the base sequence of a gene.

CHROMOSOME (p. 175)
A linear or circular strand of DNA on which are found specific sequences of base pairs. The human genome consistsof two copies of each of 23 unique chromosomes, one from the mother and one from the father. [Gk., chroma, color + soma, body]

ALLELE (p. 176)
Alternative versions of a gene. [Gk., allos, another]

VLG - 22

TRAIT (p. 176)
Any characteristic or feature of an organism, such as red petal color in a flower.

INTRON (p. 177)
A non-coding region of DNA.

GENOTYPE (p. 178)
The genes that an organism carries for a particular trait; also, collectively, an organism's genetic composition. [Gk., genos, race, descent + typos, impression, engraving]

PHENOTYPE (p. 179)
The manifested structure, function, and behaviors of an individual; the expression of the genotype of an organism. [Gk., phainein, to cause to appear + typos impression, engraving]

TRANSCRIPTION (p. 179)
The process by which a gene's base sequence is copied to mRNA.

TRANSLATION (p. 179)
The process by which mRNA, which encodes a gene's base sequence, directs the production of a protein.

MESSENGER RNA (MRNA) (p. 179)
The ribonucleic acid that "reads" the sequence for a gene in DNA and then moves from the nucleus to the cytoplasm, where the next stage of protein synthesis will take place.

PROMOTER SITE (p. 180)
Part of a DNA molecule that indicates where the sequence of base pairs that makes up a gene begins.

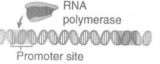

RNA polymerase

Promoter site

RIBOSOMAL SUBUNITS (p. 182)
The two structural parts of a ribosome, which function together to translate mRNA to build a chain of amino acids that will make up a protein.

TRANSFER RNA (TRNA) (p. 182)
The type of RNA molecules in the cytoplasm that link specific triplet base sequences on mRNA to specific amino acids.

CODON (p. 183)
Three-base sequences in mRNA that link with complementary tRNA molecules, which are attached to amino acids; a codon with yet another sequence ends the process of assembling a protein from amino acids.

PROTEIN SYNTHESIS (p. 185)
The construction of a protein from its constituent amino acids, by the processes of transcription and translation.

GENE EXPRESSION (p. 185)
The process by which information in a gene's sequence is used to synthesize a gene product (commonly a protein, but also RNA).

GENE REGULATION (p. 185)
The processes by which cells "turn on" or "turn off " genes, influencing the amount of gene products formed.

OPERON (p. 187)
A group of several genes, along with the elements that control their expression as a unit, all within one section of DNA.

MUTATION (p. 189)
An alteration in the base-pair sequence of an individual's DNA; may arise spontaneously or following exposure to a mutagen. [Lat., mutare, to change]

POINT MUTATION (p. 190)
A mutation in which one base pair in DNA is replaced with another, or a base pair is either inserted or deleted.

CHROMOSOMAL ABERRATION (p. 190)
A type of mutation characterized by a change in the overall organization of genes on a chromosome, such as the deletion of a section of DNA; the moving of a gene from one part of a chromosome to elsewhere on the same chromosome or to a different chromosome; or the duplication of a gene, with the new copy inserted elsewhere on the chromosome or on a different chromosome. [Lat., aberrare, to wander]

BIOTECHNOLOGY (p. 194)
The modification of organisms, cells, and their molecules for practical benefits. [Gk., bios, life + technologia, systematic treatment]

GENETIC ENGINEERING (p. 194)
The manipulation of an organism's genetic material by adding, deleting, or transplanting genes from one organism to another.

RESTRICTION ENZYMES (p. 195)
Enzymes that recognize and bind to different specific sequences of four to eight bases in DNA and cut the DNA at that point. Restriction enzymes are important in biotechnology because they permit the cutting of short lengthsof DNA, which can be inserted into other chromosomes or otherwise utilized.

POLYMERASE CHAIN REACTION (PCR) (p. 195)
A laboratory technique in which a fragment of DNA can be duplicated repeatedly. [Gk., polys, many + meris, part]

TRANSGENIC ORGANISM (p. 196)
An organism that contains DNA from another species.

CLONING (p. 196)
The production of genetically identical cells, organisms, or DNA molecules.

PLASMID (p. 196)
A circular DNA molecule found outside the main chromosome in bacteria.

CLONE (p. 196)
A genetically identical DNA fragment, cell, or organism produced by a single cell or organism. [Gk., klon, twig]

CLONE LIBRARY (OR GENE LIBRARY) (p. 197)
A collection of cloned DNA fragments; also known as a gene library.

DNA PROBE (p. 198)
A short sequence of radioactively tagged singlestranded DNA that contains part of the sequence of the gene of interest, used to locate that gene in a gene library. The probe binds to the complementary base pair on a gene in the library, which is identified by the radioactive tag on the probe.

RECOMBINANT DNA TECHNOLOGY (p. 198)
Technology that depends on the combination of two or more sources of DNA into a product; an example is the production of human insulin from fastdividing transgenic E. coli bacteria in which has been inserted the human DNA sequence that codes for insulin.

GENE THERAPY (p. 208)
The process of inserting, altering, or deleting one or more of the genes in an individual's cells to correct defective versions of the gene(s).

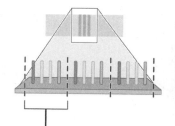

STEM CELL (p. 208)
Undifferentiated cells that have the ability to develop into any type of cell in the body; this property makes stem cells useful in biotechnology.

SHORT TANDEM REPEATS (STRS) (p. 212)
Region in DNA in which a short sequence (commonly four or five nucleotides) repeats over and over, a dozen or more times. Variation among individuals in the number of times the sequence repeats make it possible to use these regions as a DNA "fingerprint," unique to an individual and valuable as a forensic tool.

VISUAL LEARNING GLOSSARY: Key Terms in the order you see them in the text.

TELOMERE (p. 222)

A non-coding, highly repetitive section of DNA at the tip of every eukaryotic chromosome that shortens with every cell division; if it becomes too short, additional cell division can cause the loss of functional, essential DNA and therefore almost certain cell death. [Gk., telos, end + meris, part]

HISTONE (p. 224)

Proteins around which the long, linear strands of DNA are wrapped; serve to keep the DNA untangled and to enable an orderly, tight, and efficient packing of the DNA within the cell.

BINARY FISSION (p. 225)

A type of asexual reproduction in which the parent cell divides into two genetically identical daughter cells. Bacteria and other prokaryotes reproduce by binary fission. [Lat., binarius, consisting of two + fissus, divided]

REPLICATION (p. 225)

The process in both eukaryotes and prokaryotes by which DNA duplicates itself in preparation for cell division.

PARENT CELL (p. 225)

Cells that divide to form daughter cells, which are genetically identical to the parent cell.

DAUGHTER CELL (p. 225)

Cells produced by the division of a parent cell.

ASEXUAL REPRODUCTION (p. 225)

A type of reproduction common in prokaryotes and plants, and also occurring in many other multicellular organisms, in which the offspring inherit their DNA from a single parent.

SEXUAL REPRODUCTION (p. 225)

A type of reproduction in which offspring are produced by the fusion of gametes from two distinct sexes.

CELL CYCLE (p. 226)

In a cell, the alternation of activities related to cell division and those related to growth and metabolism.

SOMATIC CELL (p. 226)

The (usually diploid) cells of the body of an organism (in contrast to the usually haploid reproductive cells). [Gk., soma, body]

REPRODUCTIVE CELL (p. 226)
Haploid cells from two individuals that, as sperm and egg, will combine at fertilization to produce offspring; also called gametes.

GAMETE (p. 226)
Cell (often haploid) that will combine at fertilization to produceoffspring; also called a reproductive cell. [Gk., gamete, wife]

INTERPHASE (p. 226)
In the cell cycle, the phase during which the cell grows and functions; during this phase, replication of DNA occurs in preparation for celldivision. [Lat., inter, between + Gk., phasis, appearance]

MITOTIC PHASE (M PHASE) (p. 226)
The phase of the cell cycle during which first the genetic material and nucleus, and then the rest of the cellular contents, divide.

M

CENTROMERE (p. 227)
After replication, the region of contact between sister chromatids, which occurs near the center of the two strands. [Gk., centron, the stationary point of a pair of compasses, thus the center of a circle + meris, part]

MITOSIS (p. 227)
The division of a nucleus into two genetically identical nuclei that, along with cytokinesis, leads to the formation of two identical daughter cells. [Gk., mitos, thread + phasis, appearance]

CYTOKINESIS (p. 227)
In the cell cycle, the stage following mitosis in which cytoplasm and organelles duplicate and are divided into approximately equal parts and the cell separates into two daughter cells. In meiosis, two diploid daughter cells are formed in cytokinesis following telophase I and four haploid daughter cells are formed in cytokinesis following telophase II. [Gk., kytos, container + kinesis, motion]

COMPLEMENTARITY (p. 227)
The characteristic of double-stranded DNA that the base on one strand always has the same pairing partner, or complementary base, on the other strand.

COMPLEMENTARY BASE (p. 227)
A base on a strand of double-stranded DNA that is a pairing partner to a base on the other strand: adenine (A) is the complementary base to thymine (T), and guanine (G) is the complementary base to cytosine (C).

APOPTOSIS (p. 230)
Programmed cell death, which takes place particularly in parts of the body where the cells are likely to accumulate significant genetic damage over time and are therefore at high risk of becoming cancerous. [Gk., apoptosis, falling away]

PROPHASE (p. 232)

The first phase of mitosis, in which the nuclear membrane breaks down, sister chromatids condense,  and the spindle forms. In meiosis, homologous pairs of sister chromatids come together and cross over in prophase I, and the chromosomes in daughter cells condense in prophase II. [Gk., pro, before + phasis, appearance]

CHROMATID (p. 233)

One of the two strands of a replicated chromosome. [Gk., chroma, color]

SISTER CHROMATIDS (p. 233)

The two identical strands of a replicated chromosome.

SPINDLE (p. 233)

A part of the cytoskeleton of a cell, formed in prophase (in mitosis) or in prophase I (in meiosis), from which extend the fibers that organize and separate the sister chromatids.

CENTRIOLE (p. 233)

A structure, located outside the nucleus in most animal cells, to which the spindle fibers are attached during cell division.

SPINDLE FIBER (p. 233)

Fibers that extend from one pole of a cell to the other, which pull the sister chromatids apart in the anaphase stage of mitosis or the anaphase II stage of meiosis.

METAPHASE (p. 233)

The second phase of mitosis, in which the sister chromatids line up at the center of the cell; in meiosis, the homologues line up at the center of the cell in metaphase I and the sister chromatids line up in metaphase II. [Gk., meta, in the midst of + phasis, appearance]

ANAPHASE (p. 233)

A phase in mitosis and meiosis in which chromosomes separate. In mitosis, it is the third phase, in which the sister chromatids are pulled apart by the spindle fibers, with a full set of chromosomes going to opposite sides of the cell. [Gk., ana, up + phasis, appearance]

TELOPHASE (p. 234)

The fourth and last phase of mitosis, in which the chromosomes begin to uncoil and the nuclear membrane is reassembled around them. In meiosis, the sister chromatids arrive at the cell poles and the nuclear membrane reassembles around them in telophase I; in telophase II, the sister chromatids have been pulled apart and the nuclear membrane reassembles around haploid numbers of chromosomes. [Gk., telos, end+ phasis, appearance]

CANCER (p. 234)

Unrestrained cell growth and division. [Lat., cancer, crab]

FERTILIZATION (p. 237)

The fusion of two reproductive cells.

MEIOSIS (p. 237)
In sexually reproducing organisms, a process of nuclear division in the gonads that, along with cytokinesis, produces reproductive cells that have half as much genetic material as the parent cell and that all differ from each other genetically. [Gk., meioun, to lessen]

DIPLOID (p. 237)
Describes cells that have two copies of each chromosome(in many organisms, including humans, somatic cells are diploid). [Gk., diplasiazein, to double]

HAPLOID (p. 237)
Describes cells that have a single copy of each chromosome(in many species, including humans, gametes are haploid). [Gk.,haploeides, single]

GONAD (p. 239)
The ovaries and testes in sexually reproducing animals. [Gk., gone, offspring]

HOMOLOGOUS PAIR (HOMOLOGUES) (p. 239)
The maternal and paternal copies of a chromosome. [Gk., homologia,agreement]

RECOMBINATION (OR CROSSING OVER)(p. 241)
The exchange of some genetic material between a paternal homologous chromosome and a maternal homologous chromosome, leading to a chromosome carrying genetic material from each; also referred to as crossing over.

POLAR BODY (p. 243)
One of the two cells formed when a primary oocyte divides; it gets almost no cytoplasm and eventually disintegrates.

X AND Y CHROMOSOMES, (p. 247)
The human sex chromosomes.

HERMAPHRODITE (p. 249)
An organism that produces both male and female gametes. [From the names of the Greek god Hermes and goddess Aphrodite]

KARYOTYPE (p. 250)
A visual display of an individual's full set of chromosomes. [Gk., karyon, nut or kernel + typos, impression, engraving]

PLACENTA (p. 252)
The organ formed during pregnancy (and expelled at birth) that connects the developing embryo to the wall of the uterus and allows the transfer of gases, nutrients, and waste products between mother and fetus; the placenta is so called from its shape. [Lat., placenta, a flat cake]

NONDISJUNCTION (p. 252)
The unequal distribution of chromosomes during cell division; can lead to Down syndrome and other disorders caused by an individual's possession of too few or too many chromosomes.

VISUAL LEARNING GLOSSARY: Key Terms in the order you see them in the text.

ALLELE (p. 265)
Alternative versions of a gene. [Gk., allos, another]

HEREDITY (p. 266)
The greater resemblance of offspring to parents than to other individuals in the population, a consequence of the passing of characteristics from parents to offspring through their genes. [Lat., heres, heir]

SINGLE-GENE TRAIT (p. 267)
A trait that is determined by instructions on only one gene; examples are a cleft chin, a widow's peak, and unattached earlobes.

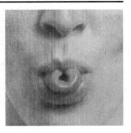

CROSS (p. 268)
The breeding of organisms that differ in one or more traits.

TRUE-BREEDING (p. 268)
Describes a population of organisms in which, for a given trait, the offspring of crosses of individuals within the population always show the same trait; for example, the offspring of pea plants that are true-breeding for round peas always have round peas.

DOMINANT (p. 270)
Describes an allele that masks the phenotypic effect of the other, recessive, allele for a trait; the phenotype shows the effect of the dominant allele in both homozygous and heterozygous genotypes. [Lat., dominari, to rule]

RECESSIVE (p. 270)
Describes an allele whose phenotypic effect is masked by a dominant allele for a trait. [Lat., recessus, retreating]

HOMOZYGOUS (p. 271)
Describes the genotype of a trait for which the two alleles are the same. [Gk., homos, same + zeugos, pair]

HETEROZYGOUS (p. 271)
Describes the genotype of a trait for which the two alleles an individual carries differ from each other. [Gk., heteros, other + zeugos, pair]

MENDEL'S LAW OF SEGREGATION (p. 271)
During the formation of gametes, the two alleles for a gene separate, so that half the gametes carry one allele, and half the gametes carry the other. [From the name of its discoverer, Gregor Mendel, 1822–1884]

PHENOTYPE (p. 272)

The manifested structure, function, and behaviors of an individual; the expression of the genotype of an organism. [Gk., phainein, to cause to appear + typos impression, engraving]

GENOTYPE (p. 272)

The genes that an organism carries for a particular trait; also, collectively, an organism's genetic composition. [Gk., genos, race, descent + typos, impression, engraving]

PUNNETT SQUARE (p. 272)

A diagram showing the possible outcomes of a cross between two individuals; the possible crosses are shown in the manner of a multiplication table. [From the name of its designer, Reginald C. Punnett, 1875–1967]

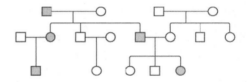

TEST-CROSS (p. 276)

A mating in which a homozygous recessive individual is bred with individuals of unknown genotype, showing the dominant phenotype; this type of cross can reveal the unknown genotype by the observed characteristics, or phenotype, of the offspring.

SEX-LINKED TRAIT (p. 277)

A trait controlled by a gene on a sex chromosome.

PEDIGREE (p. 277)

In genetics, a type of family tree that maps the occurrence of a trait in a family, often over many generations.

CARRIER (p. 278)

An individual who carries one allele for a recessive trait and does not exhibit the trait; if two carriers mate, they may produce offspring who do exhibit the trait.

INCOMPLETE DOMINANCE (p. 279)

The case in which the heterozygote has a phenotype intermediate between those of the two homozygotes; an example is pink snapdragons, with an appearance intermediate between homozygous for white flowers and homozygous for red flowers.

CODOMINANCE (p. 280)

The case in which the heterozygote displays characteristics of both alleles.

MULTIPLE ALLELISM (p. 281)

The case in which a single gene has more than two possible alleles.

POLYGENIC (p. 283)

Describes a trait that is influenced by multiple different genes. [Gk., polys, many + genos, race, descent]

ADDITIVE EFFECTS (p. 283)

Effects from alleles of multiple genes that all contribute to the ultimate phenotype for a given characteristic.

PLEIOTROPY (p. 284)

A phenomenon in which an individual gene influences multiple traits. [Gk., pleion, more + tropos, turn]

IF...
Parents are both heterozygous for both traits (i.e., "doubly heterozygous"). Four different types of gametes are produced by each: **AD**, **Ad**, **aD**, and **ad**.

MOTHER
pigmented heterozygous **Aa**
dimpled chin heterozygous **Dd**

FATHER
pigmented heterozygous **Aa**
dimpled chin heterozygous **Dd**

GAMETES

	AD	Ad	aD	ad
AD	AA DD	AA Dd	Aa DD	Aa Dd
Ad	AA Dd	AA dd	Aa Dd	Aa dd
aD	Aa DD	Aa Dd	aa DD	aa Dd
ad	Aa Dd	Aa dd	aa Dd	aa dd

MENDEL'S LAW OF INDEPENDENT ASSORTMENT (p. 290)

Allele pairs for different genes separate independently in meiosis, so the inheritance of one trait generally does not influence the inheritance of another trait (the exception, unknown to Mendel, occurs with linked genes). [From the name of its discoverer, Gregor Mendel, 1822–1884]

LINKED GENE (p. 291)

Genes that are close to each other on a chromosome and so are more likely than others to be inherited together.

8. Evolution and Natural Selection | Darwin's dangerous idea

POPULATION (p. 300)
A group of organisms of the same species living in a particular geographic region.

EVOLUTION (p. 302)
A change in allele frequencies of a population. [Lat., evolvere, to roll out]

TRAIT (p. 306)
Any characteristic or feature of an organism, such as red petal color in a flower.

MUTATION (p. 311)
An alteration in the base-pair sequence of an individual's DNA; may arise spontaneously or following exposure to a mutagen. [Lat., mutare, to change]

GENETIC DRIFT (p. 313)
A random change in allele frequencies over successive generations; a cause of evolution.

FIXATION (p. 313)
The point at which the frequency of an allele in a population is 100% and, therefore, there is no more variation in the population for this gene.

FOUNDER EFFECT (p. 314)
A phenomenon by which genetic drift can occur; the isolation of a small subgroup of a larger population which can lead to changes in the allele frequencies of a population because all the descendants of the smaller group will reflect the allele frequencies of the subgroup, which may be different from those of the larger source population.

BOTTLENECK EFFECT (p. 314)
A phenomenon by which genetic drift can occur; a sudden reduction in population size (often due to famine, disease, or rapid environmental disturbance) that can lead to changes in the allele frequencies of a population.

MIGRATION (p. 315)
A change in the allele frequencies of a population due to the movement of some individuals from one population to another; an agent of evolutionary change caused by the movement of individuals into or out of a population. [Lat., migrare, to move from place to place]

GENE FLOW (p. 315)
A change in the allele frequencies of a population due to movement of some individuals of a species from one population to another, changing the allele frequencies of the population they join; also known as migration.

NATURAL SELECTION (p. 316)

A mechanism of evolution that occurs when there is heritable variationfor a trait and individuals with one version of the trait have greater reproductive success than individuals with a different version of that trait.

INHERITANCE (OR HERITABILITY) (p. 318)

The transmission of traits from parents to offspring via genetic information; also known as heritability.

DIFFERENTIAL REPRODUCTIVE SUCCESS (p. 319)

The situation in which individuals have greater reproductive success than other individuals in a population; along with variation and heritability, differential reproductive success is one of the three conditions necessary for evolution by natural selection.

SEXUAL SELECTION (p. 319)

The process by which natural selection favors traits, such as ornaments or fighting behavior, that give an advantage to individuals of one sex in attracting mating partners.

FITNESS (p. 322)

A relative measure of the reproductive output of an individual with a given phenotype compared with the reproductive output of individuals with alternative phenotypes.

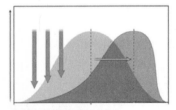

ADAPTATION (p. 323)

The process by which, as a result of natural selection, a population's organisms become better matched to their environment; also, a specific feature, such as the quills of a porcupine, that makes an organism more fit.

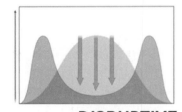

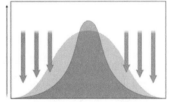

DIRECTIONAL SELECTION (p. 326)

Selection that, for a given trait, increases fitness at one extreme of the phenotype and reduces fitness at the other, leading to an increase or decrease in the mean value of the trait.

STABILIZING SELECTION (p. 327)

Selection that, for a given trait, produces the greatest fitness at the intermediate point of the phenotypic range.

DISRUPTIVE SELECTION (p. 328)

Selection that, for a given trait, increases fitness at both extremes of the phenotype distribution and reduces fitness at middle values.

FOSSIL (p. 331)

The remains of an organism, usually its hard parts such as shell, bones, or teeth, which have been naturally preserved; also, traces of such an organism, such as footprints. [Lat., fossilis, that which is dug up]

RADIOMETRIC DATING (p. 332)

A method of determining both the relative and the absolute ages of objects such as fossils by measuring both the radioactive isotopes they contain, which are known to decay at a constant rate, and their decay products.

BIOGEOGRAPHY (p. 334)

The study and interpretation of distribution patterns of living organisms around the world. [Gk. bios, life + geo-, earth + graphein, to write down]

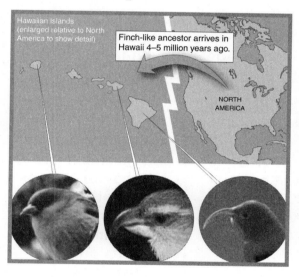

Hawaiian Islands (enlarged relative to North America to show detail)

Finch-like ancestor arrives in Hawaii 4–5 million years ago.

NORTH AMERICA

HOMOLOGOUS STRUCTURE (p. 336)

Body structures in different organisms that, although they may have been modified over time to serve different functions in different species, are due to inheritance from a common evolutionary ancestor.

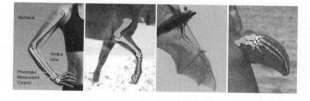

VESTIGIAL STRUCTURE (p. 336)

A structure, once useful to organisms, but which has lost its function over evolutionary time; an example is molars in bats that now consume an exclusively liquid diet. [Lat., vestigium, footprint, trace]

CONVERGENT EVOLUTION (p. 336)

A process of natural selection in which features of organisms not closely related come to resemble each other as a consequence of similar selective forces. Many marsupial and placental species resemble each other as a result of convergent evolution. [Lat., con-, together with + vergere, turn + evolvere, to roll out]

VISUAL LEARNING GLOSSARY: Key Terms in the order you see them in the text.

BEHAVIOR (p. 349)
Any and all of the actions performed by an organism, often in response to its environment or to the actions of another organism.

FIXED ACTION PATTERN (p. 351)
An innate sequence of behaviors, triggered under certain conditions, that requires no learning, does not vary, and once begun runs to completion; an example is egg-retrieval in geese.

INSTINCT (OR INNATE BEHAVIOR) (p. 351)
Behaviors that do not require environmental input for their development. These behaviors are present in all individuals in a population and do not vary much from one individual to another or over an individual's life span; also known as instincts. [Lat., instinctus, impelled, innatus, inborn]

SIGN STIMULUS (p. 351)
An external signal that triggers the innate behavior called a fixed action pattern.

LEARNING (p. 351)
The alteration and modification of behavior over time in response to experience.

PREPARED LEARNING (p. 351)
Behaviors that are learned easily by all, or nearly all, individuals of a species.

ALTRUISTIC BEHAVIOR (p. 355)
A behavior that comes at a cost to the individual performing it and benefits another. [Lat., alter, the other]

KIN SELECTION (p. 356)
"Kindness" toward close relatives, which may evolve as apparently altruistic behavior toward them, but which in fact is beneficial to the fitness of the individual performing the behavior.

RECIPROCAL ALTRUISM (p. 356)
Costly behavior directed toward another individual that benefits the recipient, with the expectation that, at some later time, the recipient will behave in a similar manner, "returning the favor." [Lat., reciprocare, to move backward and forward + alter, the other]

DIRECT FITNESS (p. 358)
The total reproductive output of an individual.

INDIRECT FITNESS (p. 358)
The reproductive output that an individual brings about through apparently altruistic behaviors toward genetic relatives.

INCLUSIVE FITNESS (p. 358)
The sum of an individual's indirect and direct fitness.

GROUP SELECTION (p. 364)
The process, extremely uncommon in nature, that brings about an increase in the frequency of alleles for traits (e.g., behaviors) that are beneficial to the persistence of the species or population while simultaneously being detrimental to the fitness of the individual possessing the trait (or engaging in the behavior).

FEMALE (p. 365)
In sexually reproducing organisms, a member of the sex that produces the larger gamete.

MALE (p. 365)
In sexually reproducing organisms, a member of the sex that produces the smaller gamete.

REPRODUCTIVE INVESTMENT (p. 366)
Energy and material expended by an individual in the growth, feeding, and care of offspring.

TOTAL REPRODUCTIVE OUTPUT (p. 366)
The lifetime number of offspring produced by an individual.

PATERNITY UNCERTAINTY (p. 367)
Describes the fact that among species with internal fertilization in the female, a male cannot be 100% certain that any offspring a female produces are his. [Lat., pater, father]

MATE GUARDING (p. 371)
Behavior by an individual that reduces the opportunity for that individual's mate to interact with other potential mates.

NUPTIAL GIFT (p. 370)
A food item or other item presented to a potential mate as part of courtship. [Lat., nuptialis, of marriage]

POLYGAMY (p. 373)
A mating system in which, for one sex, some individuals attract multiple mates while other individuals of that sex attract none; among the opposite sex, all or nearly all of the individuals are able to attract a mate. [Gk., polys, many + gamete, wife]

POLYGYNY (p. 373)
A mating system in which, among the males, some individuals attract multiple mates while other males attract none; among the females, all or nearly all of the individuals are able to attract a mate. [Gk., polys, many + gune, woman]

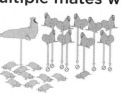

POLYANDRY (p. 373)
A polygamous mating system in which individual females mate with multiple males. [Gk., polys, many + aner, husband]

MONOGAMY (p. 373)
A mating system in which most individuals mate and remain with just one other individual. [Gk., monos, single + gamete, wife]

MATING SYSTEM (p. 373)
The pattern of mating behavior in a species, ranging from polyandry tomonogamy to polygyny; mating systems are influenced by the relative amounts of parental investment by males and by females.

PAIR BOND (p. 374)
A bond between an individual male and female in which they spend a high proportion of their time together, often over many years, sharing a nest or other refuge and contributing equally to the care of offspring.

SEXUAL DIMORPHISM (p. 375)
The case in which the sexes of a species differ in size or appearance. [Gk., di-, two + morphe, shape]

COMMUNICATION (p. 377)
An action or signal on the part of one organism that alters the behavior of another organism.

PHEROMONE (p. 377)
Molecules released by an individual into the environment that trigger behavioral or physiological responses in other individuals. [Gk., pherein, to carry]

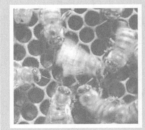

WAGGLE DANCE (p. 377)
Behavior of scout honeybees that indicates, by the angle of the body relative to the sun and by physical maneuvers of various duration, the direction to a distant source of food.

LANGUAGE (p. 378)
A type of communication in which arbitrary symbols represent concepts and grammar; a system of rules dictates how the symbols can be manipulated to communicate and express ideas.

HONEST SIGNAL (p. 379)
A signal, which cannot be faked, that is given when both the individual making the signal and the individual responding to it have the same interests; it carries the most accurate information about an individual or situation.

VISUAL LEARNING GLOSSARY: Key Terms in the order you see them in the text.

LIFE (p. 388)

A physical state characterized by the ability to replicate and the presence of metabolic activity.

BIODIVERSITY (p. 388)

The variety and variability among all genes, species, and ecosystems. [Gk., bios, life + Lat., diversus, turned in different directions]

RNA WORLD HYPOTHESIS (p. 390)

A hypothesis that proposes that the world may have been filled with RNA-based life before it became filled with life based on DNA, the lifeof today.

MICROSPHERE (p. 391)

A membrane-enclosed, small, spherical unit containing a self-replicatingmolecule and carrying information, although no genetic material. Microspheres may have been an important stage in the development of life. [Gk., micros, small + sphaira, ball]

SPECIES (p. 392)

Natural populations of organisms that can interbreed and are reproductively isolated from other such groups; in the Linnaean system, the species is the narrowest classification for an organism. [Lat.,species, kind, sort]

BIOLOGICAL SPECIES CONCEPT (p. 392)

A definition of species described as populations of organisms that interbreed, or could possibly interbreed, with each other under natural conditions, and that cannot interbreed with organisms outside their own group.

REPRODUCTIVE ISOLATION (p. 392)

The inability of individuals from two populations to produce fertile offspring together.

PREZYGOTIC BARRIER (p. 394)

A barrier to reproduction caused by the physical inability of individuals to mate with each other, or the inability of the male's reproductive cell to fertilize the female's reproductive cell. [Lat., prae, before]

POSTZYGOTIC BARRIER (p. 394)

A barrier to reproduction caused by the infertility of hybrid individuals or the inability of hybrid individuals to survive long after fertilization. [Lat., post, after]

HYBRID (p. 394)

Offspring of individuals of two different species. [Lat., hybrida, animal produced by two different species]

GENUS (p. 394)
In the system developed by Carolus Linnaeus (1707–1778), a classification of organisms consisting of closely related species. [Lat., genus, race, family, origin]

SPECIFIC EPITHET (p. 394)
In the system developed by Carolus Linnaeus (1707–1778), a noun or adjective added to the genus name to distinguish a species; in the name Homo sapiens, Homo is the genus name and sapiens is the specific epithet. [Gk., epithetos, added]

FAMILY (p. 394)
In the system developed by Carolus Linnaeus (1707–1778), a classification of organisms consisting of related genera.

ORDER (p. 394)
In the system developed by Carolus Linnaeus (1707–1778), a classification of organisms consisting of related families.

CLASS (p. 394)
In the hierarchical taxonomic system developed by Carolus Linnaeus (1707–1778), a classification of organisms consisting of related orders.

PHYLUM (p. 394)
In the system developed by Carolus Linnaeus (1707– 1778), a classification of organisms consisting of related classes. [Gk., phylon, race, tribe, class]

KINGDOM (p. 394)
In the system developed by Carolus Linnaeus (1707– 1778), one of the three categories—animal, plant, and mineral—into which all organisms and substances on earth were placed. In modern classification, there are six kingdoms: bacteria, archaea, protists, lants, animals, and fungi.

DOMAIN (p. 394)
In modern classification, the highest level of the hierarchy; there are three domains, Bacteria, Archaea, and Eukarya.

RING SPECIES (p. 396)
Populations that can interbreed with neighboring populations but not with populations separated by larger geographic distances. Because the non-interbreeding populations are connected by gene flow through geographically intermediate populations, there is no clear point at which one species stops and another begins, and for this reason, ring species are problematic for the biological species concept.

HYBRIDIZATION (p. 397)
The interbreeding of closely related species. [Lat., hybrida, animal produced by two different species]

MORPHOLOGICAL SPECIES CONCEPT (p. 397)
A concept that defines species on the basis of physical features such as body size and shape. [Gk., morphe, shape]

SPECIATION (p. 397)
The process by which one species splits into two distinct species; the first phase of speciation is reproductive isolation, the second is genetic divergence, in which two populations evolve over time as separate entities with physical and behavioral differences.

ALLOPATRIC SPECIATION (p. 397)
Speciation that occurs as a result of a geographic barrier between groups of individuals that leads to reproductive isolation and then genetic divergence. [Gk., allos, another + patris, native land]

SYMPATRIC SPECIATION (p. 398)
Speciation that results not from geographic isolation but from polyploidy or hybridization and allopolyploidy; this type of speciation is relatively uncommon in animals but is common among plants. [Gk., syn, together with + patris, native land]

POLYPLOIDY (p. 398)
The doubling of the number of sets of chromosomes in an individual. [Gk., polys, many + ploion, vessel]

SYSTEMATICS (p. 401)
The modern approach to classification, with the broader goal of reconstructing the evolutionary history, or phylogeny, of organisms. [Gk., systema, a whole compounded of parts]

PHYLOGENY (p. 401)
The evolutionary history of organisms.

SPECIATION EVENT (p. 401)
A point in evolutionary history at which a given population splits into independent evolutionary lineages.

NODE (p. 402)
The point on an evolutionary tree at which species diverge from a common ancestor. [Lat., nodus, knot]

MONOPHYLETIC (p. 403)
A group containing a common ancestor and all of its descendants. [Gk., monos, single + phylon, race, tribe, class]

Animals Fungi Plants

A

B

PROTIST (p. 404)
In modern classification, one of the four eukaryotic kingdoms; includevs all the single-celled eukaryotes.

CONVERGENT EVOLUTION (p. 405)
A process of natural selection in which features of organisms not closely related come to resemble each other as a consequence of similar selective forces. Many marsupial and placental species resemble each other as a result of convergent evolution. [Lat., con-, together with + vergere, turn + evolvere, to roll out]

ANALOGOUS TRAIT (p. 406)
Characteristics (such as bat wings and insect wings) that are similar because they were produced by convergent evolution, not because they descended from a common structure in a shared ancestor. [Gk.,analogos, proportionate]

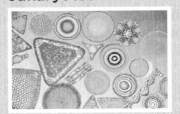

HOMOLOGOUS FEATURE (p. 406)
A feature inherited from a common ancestor. [Gk., homologia, agreement]

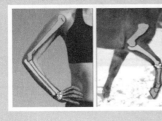

MACROEVOLUTION (p. 407)
Large-scale evolutionary change involving the origins of new groups of organisms; the accumulated effect of microevolution over a long period of time. [Gk., macros, large + Lat., evolvere, to roll out]

MICROEVOLUTION (p. 407)
A slight change in allele frequencies in a population over one or a few generations. [Gk., micros, small + Lat., evolvere, to roll out]

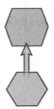

PUNCTUATED EQUILIBRIUM (p. 408)
A theory in biology about the tempo of evolution, suggesting that in many taxa long periods in which there is relatively little evolutionary change are punctuated by brief periods of rapid evolutionary change.

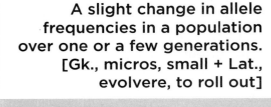

ADAPTIVE RADIATION (p. 410)
The rapid diversification of a small number of species into a much larger number of species, able to live in a wide variety of habitats.

EXTINCTION (p. 411)
The complete loss of all individuals in a species. [Lat., extinguere, to extinguish]

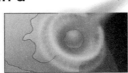

BACKGROUND EXTINCTION (p. 411)
Extinctions that occur at lower rates than at times of mass extinctions; occur mostly as the result of aspects of the biology and competitive success of the species, rather than catastrophe.

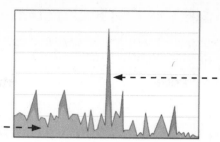

MASS EXTINCTION (p. 411)
Extinctions in which a large number of species become extinct in a short period of time, usually because of extraordinary and sudden environmental change. [Lat., extinguere, to extinguish]

MICROBE (p. 413)
A microscopic organism; not a monophyletic group, since it includes protists, archaea, and bacteria. [Gk., micros, small + bios, life]

ARCHAEA (p. 414)
A group of prokaryotes that are evolutionarily distinct from bacteria and that thrive in some of the most extreme environments on earth; one of the three domains of life. [Gk., archaios, ancient]

HORIZONTAL GENE TRANSFER (p. 414)
The transfer of genetic material directly from one individual to another, not necessarily related, individual; common among bacteria.

VIRUS (p. 414)
Diverse and important biological entities that can replicate but can conduct metabolic activity only by taking over the metabolic processes of a host organism, and therefore fall outside the definition of life. [Lat., virus, slime]

11. Animal Diversification

VISUAL LEARNING GLOSSARY: Key Terms in the order you see them in the text.

ANIMAL (p. 428)
Members of the kingdom Animalia; eukaryotic, multi-cellular, heterotrophic (that is, they cannot produce their own food) organisms. Many of these organisms have body parts specialized for different activities and can move during some stage of their lives. [Lat., animal, a living being]

SESSILE (p. 429)
Describes organisms that are fastened in place, such as adult mussels and barnacles. [Lat., sedere, to sit]
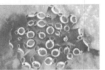

RADIAL SYMMETRY (p. 429)
A body structure like that of a wheel, or pie, in which any cut through the center would divide the organism into identical halves. [Lat., radius, spoke of a wheel + Gk., symmetria, symmetry]

BILATERAL SYMMETRY (p. 429)
A body structure with left and right sides, which are mirror images. [Lat., bi-, two + latus, side]

PROTOSTOME (p. 430)
Bilaterally symmetrical animals with defined tissues in which the gut develops from front to back; the first opening formed is the mouth of the adult animal. [Gk., protos, first + stoma, mouth]

DEUTEROSTOME (p. 430)
Bilaterally symmetrical animals with defined tissues in which the gut develops from back to front; the anus forms first, and the second opening formed is the mouth of the adult animal. [Gk., deuteros, second + stoma, mouth]

INVERTEBRATES (p. 432)
Animals that do not have a backbone; although commonly used in organizing the animals, invertebrates are not a monophyletic group.

VERTEBRATES (p. 432)
Chordate animals that have a backbone (made of cartilage or hollow bones) and, at the front end of the organism, a head containing a skull, brain, and sensory organs.

EXOSKELETON (p. 432)
A rigid external covering such as is found in some invertebrates, including insects and crustaceans. [Gk., ex, out of]

ARTHROPOD (p. 432)
Members of the invertebrate phylum Arthropoda; characterized by a segmented body, an exoskeleton, and jointed appendages. [Gk., arthron, joint + pous, foot]

FLATWORM (p. 438)

Worms with flat bodies that are members of the phylum Platyhelminthes; characterized by well-defined head and tail regions, with some having clusters of light-sensitive cells for eyespots. Most are hermaphroditic and are protostomes that do not molt. Examples are tapeworms and flukes.

ROUNDWORM (p. 438)

A worm phylum with members characterized by a long, narrow, unsegmented body and growth by molting. Roundworms, also called nematodes, are protostomes with defined tissues; there are some 90,000 identified species.

SEGMENTED WORM (p. 438)

A worm phylum with members characterized by grooves around the body that mark divisions between segments. Segmented worms, also called annelids, are protostomes with defined tissues and do not molt; examples are earthworms and leeches.

ANNELID (p. 438)

Phylum of worms having segmented bodies; protostomes with defined tissues, which grow by adding segments rather than by molting. There are about 13,000 identified species of segmented worms, including earthworms and leeches.

GASTROPOD (p. 441)

Mollusks that are members of the class Gastropoda; most have a single shell, a muscular foot for loco-motion, and a radula used for scraping food from surfaces; examples are snails and slugs. [Gk., gaster, belly + pous, foot]

BIVALVE MOLLUSK (p. 441)

Mollusks with two hinged shells; examples are clams, scallops, and oysters. [Lat., bi-, two]

CEPHALOPOD (p. 441)

Mollusks in which the head is prominent and the foot has been modified into tentacles; examples are octopuses and squids. Cephalopods have a reduced or absent shell and possess the most advanced nervous system of the invertebrates. [Gk., kephale, head + pous, foot]

LARVA (p. 446)

In complete meta-morphosis, the first stage of insect development; the larva is hatched from the egg and eats to grow large enough to enter the pupa stage. The larva (for example, a caterpillar) looks completely different from the adult (a butterfly or moth). [Lat., larva, ghost]

PUPA (p. 446)

In complete metamorphosis, the second stage of insect development, in which the larva is enclosed in a case and its body structures are broken down into molecules that are reassembled into the adult form. [Lat., pupa, a little girl]

METAMORPHOSIS (p. 446)

The rebuilding of molecules from the larva stage to the adult, resulting in a change of form. Complete metamorphosis is the division of an organism's life history into three completely different stages; incomplete metamorphosis is the pattern of growth and development in which an organism does not pass through separate, dramatically different life stages. [Gk., metamorphoun, to transform]

ADULT (p. 446)

(Pertaining to insect development stages) In complete metamorphosis, the third and final stage of insect development.

NOTOCHORD (p. 450)

A rod of tissue from head to tail that stiffens the body when muscles contract during locomotion. Primitive chordates retain the notochord throughout life, but in advanced chordates it is present only in early embryos and is replaced by the vertebral column. [Gk., notos, back + Lat., chorda, cord]

DORSAL HOLLOW NERVE CORD (p. 450)

The central nervous system of vertebrates, consisting of the spinal cord and brain. [Lat., dorsum, back]

PHARYNGEAL SLITS (p. 450)

Slits in the pharyngeal region, between the back of the mouth and the top of the throat, for the passage of water for breathing and feeding. [Gk., pharynx, throat]

POST-ANAL TAIL (p. 450)

A tail that extends beyond the end of the trunk, a point that is marked by the anus; a characteristic of chordates. [Lat., post, after]

CARTILAGINOUS FISHES (p. 454)

Fish species characterized by a skeleton made completely of cartilage, not bone. [Lat., cartilago, gristle]

RAY-FINNED FISHES (p. 454)

Fish species characterized by rigid bones and a mouth at the apex of the body; they are so called because their fins are lined with hardened rays.

LOBE-FINNED FISHES (p. 454)

Fish species characterized by two pairs of sturdy lobe-shaped fins on the underside of the body.

TETRAPOD (p. 456)

An organism with four limbs; all terrestrial vertebrates are tetrapods. [Gk., tetra-, four + pous, foot]

NON-AMNIOTES (p. 456)
Animals such as amphibians that reproduce in water and do not have desiccation-proof amniotic eggs.

AMNIOTES (p. 456)
Terrestrial vertebrates—reptiles, birds, and mammals— that produce eggs (called amniotic eggs) that are protected by a water-tight membrane and a shell.

AMPHIBIAN (p. 456)

Members of the class Amphibia; ectotherms (that is, they are coldblooded), with a moist skin, lacking scales, through which they can fully or partially absorb oxygen. They were the first terrestrial vertebrates. The young of most species are aquatic, and the adults are true land animals. [Gk., bios, life + amphi, on both sides]

ENDOTHERM (p. 458)
Organisms that use the heat produced by their cellular respiration to raise and maintain their body temperature above air temperature. [Gk., endon, within + therme, heat]

ECTOTHERM (p. 458)
Organisms that rely on the heat from an external source to raise their body temperature and seek the shade when the air is too warm. [Gk., ektos, outside + therme, heat]

HAIR (p. 459)

Dead cells filled with the protein keratin that collectively serve as insulation covering the body or a part of the body; present in all mammals.

MAMMARY GLANDS (p. 459)
Glands in all female mammals that produce milk for the nursing of young. [Lat., mamma, breast]

VIVIPARITY (p. 460)
The characteristic of bearing young alive, giving birth to babies (rather than laying eggs). [Lat., vivus, alive + parere, to bear]

MONOTREME (p. 460)
Present-day mammals that retain the ancestral condition of laying eggs. Monotremes are so called because they have a single duct, the cloaca, into which the reproductive system, the urinary system, and the digestive system (for defecation) open. [Gk., monos, single + trema, hole]

MARSUPIAL (p. 460)

Mammals in which, in most species, after a short period of embryonic life in the uterus, the young complete their development in a pouch in the female. [Gk., marsipos, pouch]

PLACENTAL (p. 460)
Mammals in which the developing fetus takes its nourishment from the transfer of nutrients from the mother through the placenta, which also supplies respiratory gases and removes metabolic waste products. [Lat., placenta, a flat cake]

VISUAL LEARNING GLOSSARY: Key Terms in the order you see them in the text.

PLANT (p. 474)
Members of the kingdom Plantae; multicellular eukaryotes that have cell walls made up primarily of cellulose, contain true tissues, and produce their own food by photosynthesis. Plants are sessile, and most inhabit terrestrial environments.

ROOT (p. 475)

The part of a vascular plant, usually below ground, that absorbs water and minerals from the soil and transports them through vascular tissue to the rest of the plant, and that anchors the plant in place. The overall vstructure of a plant's roots is called the root system.

SHOOT (p. 475)
The above-ground part of a plant, consisting of stems and leaves, and sometimes flowers and fruits. The stem contains vascular tissue and supports the leaves, the main photosynthetic organ of the plant. Also called shoot system.

NON-VASCULAR PLANT (p. 476)
Plants that do not have vessels to transport water and dissolved nutrients, but instead rely on diffusion; bryophytes are non-vascular plants. [Lat., vasculum, dim. of vas, vessel]

CUTICLE (p. 478)
A waxy layer produced by epidermal cells and found on leaves and shoots of terrestrial plants, protecting them from drying out. [Lat., cuticula, dim. of cutis, skin]

BRYOPHYTE (p. 478)
Three groups of plants (the liverworts, hornworts, and mosses) that lack vascular tissue and move water and dissolved nutrients by diffusion. [Gk., bruon, tree-moss, liverwort + phytas, plant]

GAMETOPHYTE (p. 479)
The structure in land plants and some algae that produces gametes (sperm and eggs); the haploid life stage of plants and some algae, which may be either male (producing sperm) or female (producing eggs). [Gk., gamete, wife + phytas, plant]

SPOROPHYTE (p. 479)
The multicellular diploid structure in non-vascular plants, some vascular plants, and some algae that produces asexual spores, the diploid life stage in organisms exhibiting alternation of generations.

SPORE (p. 479)

A reproductive structure of non-vascular and some vascular plants that have an alternation of generations; spores are typically haploid, unicellular, and develop into either a male (producing sperm) or female (producing eggs) gametophyte. The eggs and sperm produced by gametophytes unite to produce the diploid generation (sporophyte) of the plant. [Gk., spora, seed]

VASCULAR PLANT (p. 481)

Plants that transport water and dissolved nutrients by means of vascular tissue, a system of tubes that extends from the roots through the stem and into the leaves. [Lat., vasculum, dim. Of vas, vessel]

SPORANGIUM (p. 481)

In many ferns, the structures on the underside of the leaves in which the spores are produced. [Gk., spora, seed + angeion, vessel]

PROTHALLUS (p. 481)

The free-living haploid life stage of a fern; produces haploid gametes. [Gk., pro, before + thallia, twig]

SEED (p. 483)

An embryonic plant with its own supply of water and nutrients, encased within a protective coating.

GYMNOSPERM (p. 484)

Vascular plants that do not produce their seeds in a protective structure; seeds are usually found on the surface of the scales of a cone-like structure. The gymnosperms include conifers, cycads, gnetophytes, and ginkgo. [Gk., gymnos, naked + sperma, seed]

ENDOSPERM (p. 484)

Tissue of a mature seed that stores certain carbohydrates, proteins, and lipids that fuel the germination, growth, and development of the embryo and young seedling. [Gk., endon, within + sperma, seed]

ANGIOSPERM (p. 484)

Vascular, seed-producing flowering and fruit-bearing plants, in which the seeds are enclosed in an ovule within the ovary. [Gk., angeion, vessel, jar + sperma, seed]

POLLEN GRAIN (p. 484)

A structure that contains the male gametophyte of a seed plant. [Lat., pollen, fine dust]

OVULE (p. 484)

The structure within the ovary of flowering plants that gives rise to female egg cells. [Lat., ovum, egg]

POLLINATION (p. 484)

The transfer of pollen from the anther of one flower to the stigma of another flower. [Lat. pollen, fine dust]

FLOWER (p. 488)
The part of an angiosperm that contains the reproductive structures; consists of a supporting stem with modified leaves (the petals and sepals) and usually contains both male and female reproductive structures.

STAMEN (p. 488)
The male reproductive structure of a flower, consisting of a head-like anther on a stalk-like filament. There are usually several stamens in a flower. [Lat., stamen, thread]

ANTHER (p. 488)
The part of the stamen, the male reproductive structure of a flower, that produces pollen. [Gk., anthos, blossom]

FILAMENT (p. 488)
The supporting stalk of the anther of a stamen found in angiosperm flowers. [Lat., filum, thread]

CARPEL (p. 488)
The female reproductive structure of a flower, including the stigma, style, and ovary. [Gk., karpos, fruit]

OVARY (p. 488)
An enclosed chamber at the base of the carpel of a flower that contains the ovules; the female gonad. [Lat., ovum, egg]

DOUBLE FERTILIZATION (p. 492)
In angiosperms, two sperm are released by a pollen grain: one fuses with an egg to form a zygote, and the other fuses with two nuclei, forming a triploid endosperm.

FLESHY FRUIT (p. 494)
A fruit that consists of the ovary and some additional parts of the flower; when fleshy fruits, an attractive food, are eaten by animals, the seeds may be widely dispersed.

HYPHAE (p. 499)
(pron. HIGH-fee) Long strings of cells that make up the mycelium of a multicellular fungus. [Gk., hypha, web]

MYCORRHIZAE (p. 502)
(pron. my-ko-RYE-zay) Root fungi, that is, symbiotic associations between roots and fungi in which fungal structures are closely associated with fine rootlets and root hairs. [Gk., mykes, fungus + rhiza, root]

MYCELIUM (p. 500)
A mass of interconnecting hyphae that make up the structure of a multicellular fungus. [Gk., mykes, fungus]

DECOMPOSER (p. 501)
Organisms, including bacteria, fungi, and detritivores, that break down and feed on once-living organisms.

LICHEN (p. 503)
Symbiotic partnership between fungi and chloro-phyll-containing algae or cyanobac-teria, or both.

VISUAL LEARNING GLOSSARY: Key Terms in the order you see them in the text.

MICROBE (p. 512)
A microscopic organism; not a monophyletic group, since it includes protists, archaea, and bacteria. [Gk., micros, small + bios, life]

PEPTIDOGLYCAN (p. 516)
A glycoprotein that forms a thick layer on the outside of the cell wall of a bacterium; in some bacteria, the layer of peptidoglycan is covered by a membrane and so is not colored by a Gram stain.

CAPSULE (p. 517)
A layer surrounding the cell wall of many bacteria; it may restrict the movement of water out of the cell and thus allow bacteria to live in dry places, such as the surface of the skin. The capsule contributes to the virulent characteristics of some bacteria, making them resistant to phagocytosis by the host's immune system. [Lat., capsula, small box or case]

GRAM STAIN (p. 516)
A test used by microbiologists in identifying an unknown bacterium; the dye that is used stains the layer of peptidoglycan outside the cell wall purple, but those bacteria in which the layer of peptidoglycan is covered by a membrane are not colored by the dye. Bacteria that take a Gram stain are known as Gram-positive bacteria, those that do not are known as Gram-negative bacteria.

PLASMID (p. 518)
A circular DNA molecule found outside the main chromosome in bacteria.

TRANSDUCTION (p. 519)
A method of lateral transfer of DNA from one bacterial cell to another by means of a virus containing pieces of bacterial DNA picked up from its previous host, which infects the recipient bacterium and passes along new genes to the recipient. [Lat., trans, on the other side of + ducere, to lead]

CONJUGATION (p. 518)
The process by which a bacterium transfers a copy of some or all of its DNA to another bacterium, of the same or another species. [Lat., coniugatio, connection]

TRANSFORMATION (p. 519)
A method of lateral transfer of DNA from one bacterial cell to another in which a bacterial cell scavenges DNA released from burst bacterial cells in the environment. [Lat., trans, on the other side of + formare, to shape]

CHEMOORGANOTROPH (p. 519)

Bacteria that consume organic molecules, such as carbohydrates, as an energy source. [Gk., trophe, food]

CHEMOLITHOTROPH (p. 520)

Bacteria that can use inorganic molecules such as ammonia, hydrogen sulfide, hydrogen, and iron as sources of energy. [Gk., lithos, stone, rock + trophe, food]

PHOTOAUTOTROPH (p. 520)

Chlorophyll-containing bacteria, or other organisms, that use the energy from sunlight to convert carbon dioxide to glucose by photosynthesis. [Gk., phos, light + autos, self + trophe, food]

OXYGEN REVOLUTION (p. 520)

The accumulation in the atmosphere of oxygen released by cyanobacteria and other photosynthetic organisms.

PATHOGENIC (p. 522)

Disease-causing.

PROBIOTIC THERAPY (p. 521)

A method of treating infections by introducing benign bacteria in numbers large enough to overwhelm harmful bacteria in the body. [Lat., pro, for, on behalf of + Gk., bios, life]

SEXUALLY TRANSMITTED DISEASE (STD) (p. 525)

A disease passed from one person to another through sexual activity.

EXTREMOPHILE (p. 527)

Bacteria and archaea that can live in extreme physical and chemical conditions. [Lat., extremus, outermost + Gk., philios, loving]

PHAGOCYTOSIS (p. 530)

One of the three types of endocytosis, in which relatively large solid particles are engulfed by the plasma membrane, a vesicle is formed, and the particle is moved into the cell.

PARASITE (p. 532)

An organism that lives in or on another organism, the host, and damages it. [Gk., para, beside + sitos, grain, food]

HOST (p. 532)

An organism in or on which a parasite lives.

CAPSID (p. 533)

The protein container surrounding the genetic material (DNA or RNA) of a virus. [Lat., capsa, box, case]

ACQUIRED IMMUNO-DEFICIENCY SYNDROME (AIDS) (p. 538)

Infectious human disease caused by a retrovirus, HIV (human immuno-deficiency virus), which compromises the immune system by attacking T cells, leaving an individual susceptible to infections, as well as cancers.

HUMAN IMMUNO-DEFICIENCY VIRUS (HIV) (p. 538)

The virus responsible for AIDS, a deadly disease that destroys the human immune system. HIV is a retrovirus, an RNA-containing virus that is thought to have been introduced to humans from chimpanzees.

RETROVIRUS (p. 538)

A virus containing RNA and also reverse transcriptase, a viral enzyme, which uses the viral RNA as a template to synthesize a single strand of DNA. [Lat., retro, backward + virus, slime]

14. Population Ecology | Planet at capacity: patterns of population growth

ECOLOGY (p. 548)
The study of the interaction between organisms and their environments, at the level of individuals, populations, communities, and ecosystems. [Gk., oikos, home + logos, discourse]

POPULATION ECOLOGY (p. 549)
A subfield of ecology that studies the interactions between populations of organisms of a species and their environment.

GROWTH RATE (p. 551)
The birth rate minus the death rate; the change in the number of individuals in a population per unit of time.

EXPONENTIAL GROWTH (p. 551)
Growth of a population at a rate that is proportional to its current size.

POPULATION DENSITY (p. 552)
The number of individuals of a population in a given area.

DENSITY-DEPENDENT FACTORS (p. 552)
Limitations on a population's growth that are a consequence of population density.

CARRYING CAPACITY, K (p. 552)
The ceiling on a population's growth imposed by the limitation of resources for a particular habitat over a period of time.

LOGISTIC GROWTH (p. 553)
A pattern of population growth in which initially exponential growth levels off as the environment's carrying capacity is approached.

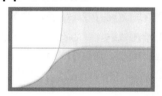

DENSITY-INDEPENDENT FACTORS (p. 553)
Limitations on a population's growth without regard to population size, such as floods, earthquakes, fires, and lightning.

MAXIMUM SUSTAINABLE YIELD (p. 557)
The point at which the maximum number of individuals are removed from a population without impairing its growth rate; it occurs at half the carrying capacity.

LIFE HISTORY (p. 560)
The vital statistics of a species, including age at first reproduction, probabilities of survival and reproduction at each age, litter size and frequency, and longevity.

REPRODUCTIVE INVESTMENT (p. 560)
Energy and material expended by an individual in the growth, feeding, and care of offspring.

LIFE TABLE (p. 561)
A table presenting data on the mortality rates within defined age ranges for a population; used to determine an individual's probability of dying during any particular year.

SURVIVORSHIP CURVE (p. 561)
Graphs showing the proportion of individuals of particular ages now alive in a population; indicate an individual's likelihood of surviving through a given age interval.

AGING (p. 564)
An increased risk of mortality with increasing age; generally characterized by multiple physiological breakdowns.

HAZARD FACTOR (p. 567)
An external force on a population that increases the risk of death.

DEMOGRAPHIC TRANSITION (p. 573)
A pattern of population growth characterized by the progression from high birth and death rates (slow growth) to high birth rates and low death rates (fast growth) to low birth and death rates (slow growth).

ECOLOGICAL FOOTPRINT (p. 575)
A measure of the impact of an individual or population on the environment by calculation of the amount of resources—including land, food and water, and fuel—consumed.

REPRODUCTIVE OUTPUT (p. 563)
The number of offspring an individual or population produces.

VISUAL LEARNING GLOSSARY: Key Terms in the order you see them in the text.

ECOSYSTEM (p. 584)
A community of biological organisms and the nonliving environmental components with which they interact.

COMMUNITY (p. 585)
The biotic environment; a geographic area defined as a loose assemblage of species with overlapping ranges.

HABITAT (p. 585)
The physical environment of organisms, consisting of the chemical resources of the soil, water, and air, and physical conditions such as temperature, salinity, humidity, and energy sources. [Lat., habitare, to dwell or inhabit]

PRIMARY PRODUCTIVITY (p. 587)
The amount of organic matter produced by living organisms, primarily through photosynthesis.

PRIMARY PRODUCER (p. 587)
The organisms responsible for primary productivity, such as grasses, trees, and agricultural crops, which convert light energy from the sun into chemical energy (that is, food) through photosynthesis.

BIOTIC (p. 585)
Relating to living organisms; the biotic environment, or community, consists of all the living organisms in a given area. [Gk., bios, life]

ABIOTIC (p. 585)
Relating to the physical and chemical components in an environment. These include the chemical resources of the soil, water, and air (such as carbon, nitrogen, and phosphorus) and physical conditions (such as temperature, salinity, moisture, humidity, and energy sources).

BIOME (p. 586)

The major ecological communities of earth. Terrestrial biomes, such as rain forest or desert, are defined and usually described by the predominant types of plant life in the area, which are mostly determined by the weather; aquatic biomes are usually defined by physical features such as salinity, water movement, and depth. [Gk., bios, life]

TROPICAL RAIN FOREST (p. 587)
A type of terrestrial biome, found between the Tropic of Cancer (23.5° north latitude) and Tropic of Capricorn (23.5° south latitude) and characterized by constant moisture and temperature that do not vary across the seasons; vegetation is dense.

SAVANNA (p. 587)
A type of terrestrial biome; a tropical or subtropical grassland with scattered woody plants, characterized by hot climate and distinct wet and dry seasons (rainfall is less than in the tropical seasonal forest biome).

TROPICAL SEASONAL FOREST (p. 587)
A type of terrestrial biome, characterized by hot climate and distinct wet and dry seasons; trees shed their leaves in the dry season.

TEMPERATE GRASSLAND (p. 553)
A type of terrestrial biome; a dry area with a hot season and a cold season (climatic conditions are less extreme than in the desert biome); vegetation is mostly grassland and shrubs.

DESERT (p. 587)
A type of terrestrial biome; a type of dry climate, with very little rainfall, in which water loss through evaporation exceeds water gain through precipitation; typically found at 30° north and south latitude.

ESTUARY (p. 587)
A tidal water passage, linked to the sea, in which salt water and fresh water mix; characterized by exceptionally high productivity. [Lat., aestus, tide]

TOPOGRAPHY (p. 590)
The physical features of a region, including those created by humans. [Gk., topos, place + graphein, to write down]

RAIN SHADOW (p. 591)
An area in the lee of a mountain where there is no or reduced rainfall, because the air passing over the mountain falls, becoming warmer and thus increasing the amount of moisture it can hold.

EL NIÑO (p. 592)
A sustained surface temperature change in the central Pacific Ocean that occurs every two to seven years; this event can start a chain reaction of unusual weather across the globe that can result in flooding, droughts, famine, and a variety of extreme climate disruptions. [Spanish, the child; a reference to the Christ child, because of the appearance of the phenomenon at Christmastime]

LA NIÑA (p. 593)
The counterpart of the El Niño phenomenon, characterized by the reverse effects, including a decrease in surface temperature and air pressure at the surface of the western Pacific ocean; the resulting climate effects are approximately opposite to El Niño effects.

PRIMARY CONSUMER (p. 594)
Herbivores, which consume the output of primary producers.

HERBIVORE (p. 594)
Animals that eat plants; also known as primary consumers. [Lat., herba, grass + vorare, to devour]

CARNIVORE (p. 594)
Predatory animals (and some plants) that consume only animals. [Lat., carnis, of flesh + vorare, to devour]

SECONDARY CONSUMER (p. 594)
Animals that feed on herbivores; also known as carnivores.

TERTIARY CONSUMER (p. 594)
Animals that eat animals that eat herbivores; also known as top carnivores. [Lat., tertius, third]

FOOD CHAIN (p. 595)
The path of energy flow from primary producers to tertiary consumers.

FOOD WEB (p. 595)
A more precisely described path of energy flow from primaryproducers to tertiary consumers than the food chain, reflecting the fact that many organisms are omnivores and occupy more than one position in the chain.

OMNIVORE (p. 595)
Animals that eat both plants and other animals and thus can occupy more than one position in the food chain. [Lat., omnis, all + vorare, to devour]

DECOMPOSER (p. 595)
Organisms, including bacteria, fungi, and detritivores, that break down and feed on once-living organisms.

DETRITIVORE (p. 595)
Organisms that break down and feed on once-living organic matter; this group includes scavengers such as vultures, worms, and a variety of arthropods. [Lat., detritus, worn out + vorare, to devour]

BIOMASS (p. 596)
The total mass of all the living organisms in a given area. [Gk., bios, life]

ENERGY PYRAMID (p. 596)
A diagram that illustrates the path of energy through the organisms of an ecosystem; each layer of the pyramid represents the biomass of a trophic level.

EUTROPHICATION (p. 601)

The process in which excess nutrients dissolved in a body of water lead to rapid growth of algae and bacteria, which consume much of the dissolved oxygen and, in time, can lead to large-scale die-offs. [Gk., eu, good + trophe, food]

COEVOLUTION (p. 602)

The concurrent appearance and modification over time, through natural selection, of traits in interacting species that enable each species to become adapted to the other; an example is the 11-inch-long tongue of a moth that feeds from the 11-inch-long nectar tube of an orchid.

REALIZED NICHE (p. 603)

The environmental conditions in which an organism is living at a given time.

COMPETITIVE EXCLUSION (p. 604)

The case in which two species battle for resources in the same niche until the more efficient of the two wins and the other is driven to extinction in that location.

CHARACTER DISPLACEMENT (p. 604)

An evolutionary divergence in one or more of the species that occupy a niche that leads to a partitioning of the niche between the species. Changes in characteristics, such as behavior or body plan, of two or more very similar species that have overlapping geographic locations result in a reduction of competition between the species.

NICHE (p. 603)

The way an organism utilizes the resources of its environment, including the space it requires, the food it consumes, and timing of reproduction.

FUNDAMENTAL NICHE (p. 603)

The full range of environmental conditions under which an organism can live.

RESOURCE PARTITIONING (p. 604)

A division of resources that occurs when species overlap some portion of a niche in which one or more species differ in behavior or body plan in a way that divides the resources of the niche between the species.

PREDATION (p. 605)

An interaction between two species in which one species eats the other. [Lat., praedari, to plunder]

PHYSICAL DEFENSES (p. 606)

Defenses that include mechanical, chemical, warning coloration, and camouflage mechanisms.

MIMICRY (p. 607)

The evolution of an organism to resemble another organism or object in its environment to help conceal itself from predators. [Gk., mimesis, imitation]

BEHAVIORAL DEFENSES (p. 607)

Defenses that include both seemingly passive and active behaviors: hiding or escaping, and alarm calling or fighting back.

PARASITISM (p. 609)

The evolution of an organism to resemble another organism or object in its environment to help conceal itself from predators. [Gk., mimesis,imitation]

PARASITE (p. 609)

The evolution of an organism to resemble another organism or object in its environment to help conceal itself from predators. [Gk., mimesis, imitation]

HOST (p. 609)

An organism in or on which a parasite lives.

MUTUALISM (p. 611)

A symbiotic relationship in which both species benefit and neither is harmed. [Lat., mutuus, reciprocal]

COMMENSALISM (p. 611)

A symbiotic relationship between species in which one benefits and the other neither benefits nor is harmed. [Lat., com, with + mensa, table]

SUCCESSION (p. 612)

The change in the species composition of a community over time, following a disturbance.

COLONIZER (p. 612)

Species introduced into an area that has been disturbed and is undergoing the process of either primary or secondary succession. The identity of a colonizing species varies, depending on the stage and type of succession.

DISPERSER (p. 613)

Organisms able to move away from their original home.

KEYSTONE SPECIES (p. 614)

A species that has an unusually large influence on the presence or absence of numerous other species in a community.

CLIMAX COMMUNITY (p. 613)

A stable and self-sustaining community that results from ecological succession.

VISUAL LEARNING GLOSSARY: Key Terms in the order you see them in the text.

BIODIVERSITY (p. 625)
The variety and variability among all genes, species, and ecosystems. [Gk., bios, life + Lat., diversus, turned in different directions]

BIODIVERSITY HOTSPOTS (p. 629)
Regions of the world with significant reservoirs of biodiversity that are under threat of destruction.

EXOTIC SPECIES (OR INTRODUCED SPECIES) (p. 640)
Species introduced by human activities to areas other than the species' native range.

ENDANGERED SPECIES ACT (p. 652)
A U.S. law that defines "endangered species" and is designed to protect those species from extinction.

LANDSCAPE CONSERVATION (p. 653)
The conservation of habitats and ecological processes, as well as species.

CONSERVATION BIOLOGY (p. 627)
An interdisciplinary field, drawing on ecology, economics, psychology, sociology, and political science, that studies and devises ways of preserving and protecting biodiversity and other natural resources.

ENDEMIC SPECIES (p. 631)
Describes species peculiar to a particular region and not naturally found elsewhere. [Gk., en, in + demos, the people of a country]

INVASIVE SPECIES (p. 640)
Species that are introduced and cause harm.

ENDANGERED SPECIES (p. 652)
As defined by the Endangered Species Act, species in danger of extinction throughout all or a significant portion of their range.

ANSWER KEY

Chapter 1
1. a
2. b
3. d

4. b
5. c
6. b

7. c
8. e

Chapter 2
1. b
2. d
3. b

4. a
5. b
6. a

7. a
8. d
9. c

Chapter 3
1. e
2. e
3. c
4. e

5. b
6. d
7. e
8. b

9. c
10. d

Chapter 4
1. a
2. c
3. d

4. a
5. e
6. a

7. a
8. a
9. d

Chapter 5
1. c
2. d
3. d
4. d

5. e
6. c
7. a
8. a

9. a
10. e

Chapter 6
1. a
2. d
3. c
4. d

5. d
6. d
7. b
8. e

9. e
10. a

Chapter 7
1. e
2. d
3. e

4. c
5. c
6. b

7. e
8. e
9. e

Chapter 8
1. a
2. d
3. a
4. b

5. d
6. c
7. a
8. d

9. c
10. a

Chapter 9

1. e	4. d	7. c
2. a	5. a	8. c
3. a	6. a	9. d

Chapter 10

1. c	5. e	9. c
2. a	6. a	10. c
3. b	7. d	
4. a	8. b	

Chapter 11

1. a	5. c	9. b
2. a	6. a	10. e
3. b	7. b	11. e
4. d	8. b	

Chapter 12

1. c	5. c	9. d
2. c	6. d	10. c
3. b	7. b	
4. e	8. a	

Chapter 13

1. b	4. c	7. a
2. c	5. d	8. b
3. d	6. d	9. a

Chapter 14

1. e	5. a	9. e
2. d	6. d	10. c
3. d	7. e	
4. e	8. b	

Chapter 15

1. b	5. e	9. c
2. a	6. c	10. e
3. e	7. b	
4. c	8. d	

Chapter 16

1. b	6. a	
2. a	7. c	
3. e	8. a	
4. c	9. a	
5. e	10. b